EAST
SIBERIAN
SEA

D0984664

Noril'sk

East Siberian
Lowland

Kamchatka
Peninsula

Ust'-Vilyuy

Sredne-
vilyuy   520/4.1

Yakutsk

West
Siberian
Lowland

SIBERIA

SEA OF OKHOTSK

Aldan

1220 or 1420/52.1
to 112.3

Sakhalin

320/800
Krasnoyarsk

20/300

Skovorodino

Khabarovsk

1220/1000

Svobodnyy

Angarsk Irkutsk

Chita

Raychikhinsk   Birobidzhan

Ulan-Ude

CHINA

Nakhodka

LNG tanker
Oil tanker

MONGOLIA

SEA OF JAPAN

Petroleum area   Major oil refinery   diameter in millimeters

1020/23.3 capacity in million
cubic meters per day

PIPELINES   in operation   planned or under
construction   MAJOR FIELDS

diameter in millimeters

Natural gas   —————   – – – – – –   Natural gas

1220/1000 capacity in thousand
barrels per day

Crude oil   —————   – – – – – –   Crude oil

The Enigma of Soviet Petroleum

The Enigma of Soviet Petroleum

Half-Full or Half-Empty?

MARSHALL I. GOLDMAN

1919 Professor of Economics, Wellesley College
Associate Director of the Russian Research Center, Harvard
University

London
GEORGE ALLEN & UNWIN
Boston   Sydney

First published in 1980

GEORGE ALLEN & UNWIN LTD
40 Museum Street, London WC1A 1LU

The end-paper map 'Major Soviet Petroleum Deposits, Pipeline Systems and Refineries' has been redrawn from *Prospects for Soviet Oil Production, Bulletin ER77–10270* published by the Central Intelligence Agency in April 1977.

British Library Cataloguing in Publication Data

Goldman, Marshall Irwin
    The enigma of Soviet petroleum.
    1. Petroleum industry and trade – Russia
    I. Title
    338.2'7'2820947   HD9575.R82   80-40544

    ISBN 0–04–333015–0
    ISBN 0–04–333016–9 Pbk

Set in 10 on 11 point Plantin by Computacomp (UK) Ltd, Fort William, and printed and bound in Great Britain by William Clowes (Beccles) Limited, Beccles and London

Contents

Tables

To my family
for their support, patience and love

Acknowledgements

Before embarking, it would be thoughtless of me not to thank those who have helped me over the years. First, my thanks to Wellesley College and the Class of 1919 for providing me with generous financial support during my sabbatical leave and to my colleagues and students for being so understanding about the amount of time I have spent on this project. I am also grateful to the Council for the International Exchange of Scholars for awarding me a Fulbright-Hayes Exchange Lectureship which made it possible for me to spend the fall semester of 1977 at Moscow State University. Although I was there to lecture, the visit also made it possible for me to gather some material and insights I otherwise would have lacked. Most of all I am grateful to the Russian Research Center at Harvard University for the research facilities it has provided me for over twenty years. Among others, I am indebted to Susan Gardos in the library, to my long-suffering typists, Rose di Benedetto and Mary Christopher, and to my secretary, Ingrid Sayied, who not only shared in the typing but also took over some of my other responsibilities which then made it possible for me to have more time for my research and writing. Ethan Burger's help with the proof-reading and Seth and Karla Goldman's assistance with the bibliography were an unexpected but enormous help. Finally, I want to thank my wife. In this day of never-lasting marriages, ours is thriving. Even more surprising, it endures despite the fact that she spent over a month revising my manuscript. Not only was that cause for an argument or two along the way, but it also meant that she had to put her own manuscript aside. As a family we have been through much together and that we have survived as well as we have is due more to Merle than anything else.

I

Introduction

THE ORDERED AFFLUENT SOCIETY

It really wasn't, but the world seemed to be so much simpler two decades ago. In those days everything seemed to be so predictable. The main struggle was an East–West one. True, there was some jockeying for position within each bloc; the Germans versus the French and the Japanese against everyone. Similarly the Hungarians were moving in a different direction from the East Germans, the Soviets took exception to the Czechs, and the Rumanians seemed to be out of step with the whole bloc. But essentially, East–West defined the main confrontation. Granted, the southern, less developed countries (LDCs) were making an increasing amount of noise and were hinting at a greater economic and political role for themselves, but for the most part the richer countries in both East and West paid them little heed. After all, the LDCs, despite their pretensions, were not important. As Stalin might have said, they did not have the economic troops, that is, the productive or monetary clout, requisite to command attention.

In contrast to the LDCs, conditions in the industrialized world, both East and West, seemed to be reasonably well ordered. Rapid economic growth seemed to be the rule of the day throughout. Khrushchev was boasting that the Soviet Union would overtake and surpass the United States by 1970, or at the latest by 1980. However exaggerated his claims may have been, he clearly seemed to have the economic resources at his disposal. The main challenge seemed to be to stimulate Soviet planners and workers, so they would unleash their productivity and efforts on behalf of the economy.

Nor was the situation bleak in the United States, particularly after the election of John F. Kennedy. The recession of the late 1950s was soon forgotten, and the country experienced a remarkable three or four years of prolonged economic growth, combined with price stability. Even though some of our allies in Western Europe and Asia may have enjoyed higher rates of growth, ours was impressive enough for most Americans. Moreover, everything seemed subject to rational resolution: American economists were riding high. When economic growth slowed, they altered the fiscal mix and made the proper course correction. Inflation was controlled in a similar way. Life seemed all so determinant and determinable.

What we did not appreciate was that along with a good set of economic tools and economic and political practitioners, we had an abundant and rich resource endowment. Since we had always had it, we took it for granted. Only belatedly did we appreciate how important it was to be self-sufficient. Twenty years ago, our imports amounted to 3 per cent of our GNP,[1] so that the United States was not too heavily dependent on foreign sources of supply. In the critical case of petroleum, net imports amounted to about 16 per cent of American consumption, and those sources of supply were firmly under the control of American multinational companies.[2] Not only could those companies take adequate care of American needs, but they were the prime suppliers of most of Western Europe and Japan. In contrast, by 1978 total American imports had risen to 9 per cent of our GNP. Even more important, by 1979 net American oil imports amounted to almost 50 per cent of American consumption. Moreover, today there are hardly any foreign sources of supply that can be said to be firmly under the control of American companies. For that matter, as of the late 1970s, there were hardly any sources of petroleum that were under the firm control of anyone other than the host country.

For American policy-makers, these radical upheavals posed all sorts of new challenges. In the economic sphere, this meant that the salves and ointments in the medical-economic chest which had worked so well in the 1960s could no longer be depended on to bring such miraculous relief. The fiscal and monetary potions that once brought tranquility now were as likely to bring trauma. The unemployment brought on by the sudden shrinking of the petroleum supply did not disappear when interest rates and taxes were cut or expenditures increased. Instead of an increase in employment, the more likely result was an increase in prices. No longer could most economic problems be solved by simply throwing more money at them. In the extreme case of petroleum, prices were increased, but the gas lines hardly moved. With demand elasticity of about 11 per cent, every time prices doubled there would only be an 11 per cent drop in demand. That was of some help, but not enough. Even more important, higher prices now generate scarcely any increase in the supply of petroleum, at least within the United States. After 5,000 feet the deeper drillings, stimulated by the higher prices, yield more and more gas and less and less petroleum, until after 19,000 feet the deeper drillings bring virtually only gas.[3] Thus how much petroleum is pumped turns out to be more a question of revolution, Middle East turmoil, physical limitation, and the cartel decisions of Saudi Arabia, Iran, and Kuwait than a question of how much prices are increased.

THE SCARCITY SOCIETY

John Kenneth Galbraith's landmark book, *The Affluent Society*, appeared slightly more than a decade before the typology of the problem changed. By 1973 the 'affluent society' of the OECD countries (Organization for Economic and Co-operative Development), such as Western Europe, the United States, Canada, and Japan, might just as well be described as the 'scarcity society.' In the United States, at least, the switch caught us unprepared psychologically and institutionally.

Nor were we impotent only in the economic sphere. To our dismay, we barked political and military orders and found that fewer and fewer countries snapped to attention. In some cases, as in Cuba, the government was militantly hostile to the United States and supportive of or subservient to the Soviet Union. In other instances it was enough that former allies, such as the once-compliant Iran, had decided on an independent or even unco-operative stance. The United States found itself issuing fewer and fewer injunctions to the raw material producing world, especially the oil producers.

The change was felt even more strongly in Western Europe and Japan. Having intentionally closed down their coal industries, once the chief source of their energy, most of the countries of Western Europe, except for the United Kingdom, Norway, and the Netherlands, found themselves dependent almost completely on the goodwill of the oil producers. Since the Japanese had little coal to start with, there was little change in their dependency on imports. But until the 1970s, that dependency was not regarded as a vulnerability. Rather it was the raw material producers who felt themselves subservient to the advanced manufacturing nations. Beginning in the early 1970s, however, that relationship changed fundamentally, and in some instances the intimidated, seemingly overnight, became the intimidators.

The turning point was the embargo of petroleum during the 1973 war and the fourfold price increase. Yet, as in all such episodes, the change had taken root gradually, so that the ascendancy of the raw material producers was not the result of a single event.[4] The rapid increase in the demand for energy and the failure to increase supplies at a comparable rate had made a denouement inevitable.

Not all the raw material producers benefited equally from the change in market and power relationships. Thus, bauxite and coffee producers tried to flaunt what they hoped would be their power, but to no avail. As long as there seemed to be adequate supplies of a good available relative to the demand, market restraint was impractical. There would even be times when dissension would break out within OPEC ranks, suggesting that OPEC operates more as a defensive rather than an offensive unit.

At best, it tries to sustain prices already attained by its members on the spot market. It has never posted a price not already reached in the market on its own initiative, and it has never assigned production limitation quotas. None the less, OPEC and its members' power is enormous.

THE USSR AS A RAW MATERIAL PRODUCER

Throughout this period, the main focus of attention was on the OPEC nations. The Soviet Union was virtually ignored in the early phases of this new crisis. However, as the OPEC nations and some of the other raw material producing LDCs began to raise their voices and their strength, they along with many others began to associate the Soviet Union and its East European allies with the other developed countries of the North. That might have been a proper categorization for Eastern Europe, but for the Soviet Union it was, at least in part, a misconception. It is true that the Soviet Union is the world's second largest industrial power, and it also happens that most of our attention in the West has been focused on Soviet economic problems and the inefficiencies of economic planning. We have also been distracted by concern over Soviet military strength.

What has often been ignored, however, is that while the Soviet Union does indeed fulfill the criteria as one of the strongest nations of the developed world, both economically and militarily, with a different set of examiners it could also pass as a developing nation. The Soviet Union is one of the world's largest or second largest producers of a whole range of raw materials. The same thing could be said about the United States, especially since we are the world's largest producer of grain, natural gas, and certain forms of coal, and in 1979 were the world's second largest producer of petroleum. Yet, with the exception of grain and coal, we do not produce enough of the other raw materials for our needs. Therefore, we are forced to import large quantities of a whole range of raw materials. In contrast, the Soviet Union produces so much of most raw materials that even though it consumes a substantial amount of its own products it still has plenty left over for export (bauxite, lead, molybdenum, and phosphorus are important exceptions). Furthermore, its production of most raw materials has continued to grow, which, until the late 1970s, made it possible to increase the amount available for both domestic and external needs.

The Soviet Union's raw material strength is immense. Thus, the USSR is the world's largest producer of petroleum, pig iron, steel, cement, cotton, manganese, tungsten, chromium, platinum, nickel, lead, timber, and coal, if brown coal is included in the total.[5] If the lower calorie lignite and brown coal reserves are excluded, then Soviet coal reserves are considerably smaller and equivalent to about 60 per cent of American coal reserves.[6] In addition, it is the world's second largest producer of natural gas. Nor is it likely that the Soviet Union soon will lose its lead in many of these fields. For the most part, its reserves are equally impressive. For instance, it has the largest reserves of natural gas in the world.[7] Thus, according to one Soviet geographer, the Soviet Union has 59 per cent of the world's coal reserves, 41 per cent of its iron ore, 37 per cent of its natural gas, 80 per cent of its manganese, and 54 per cent of its potassium.[8] It also has substantial deposits of apatite and asbestos. It is true that a high percentage of these reserves, particularly oil and natural gas, are in remote and climatically hostile areas. But it is also true that the Soviets are used to working under such adverse conditions.

SOVIET PETROLEUM AND THE CIA

Soviet petroleum production warrants a special note. Although it is no secret, many specialists none the less are still surprised to learn that today the world's largest producer of petroleum is the Soviet Union. It assumed the lead from the United States in 1976 and continues to maintain its position despite the addition of Alaskan oil to the American total.[9] Moreover, because the Soviet Union manages to export over 25 per cent of its petroleum, it is today the world's second largest exporter of petroleum, second only to Saudi Arabia. Until the Shah was overthrown, Iran exported more than the Soviet Union. While exports from Iran were resumed shortly after the Ayatollah Khomeini came to power, they were lower than previously. The Soviet Union continues to hold on to second place at least through 1979. How long the Soviet Union will be able to sustain these exports is obviously a key question and was the focus of a Central Intelligence Agency (CIA) report in April 1977.[10] According to this CIA finding, the Soviet Union and its East European allies, by 1985, were expected to become annual net importers of 3·5 to 4·5 million barrels a day (MBD) of petroleum (175 to 225 million tons). This also means a fall in annual production.

The CIA does acknowledge that the Soviet Union has substantial reserves. While it does not accept an industry estimate that Soviet 'proved' petroleum reserves amount to 75 billion barrels, its own estimates still leave the Soviets with a substantial quantity of energy.[11] As of January 1976, the CIA estimated that Soviet proved reserves of petroleum were between 30 to 35 billion barrels (4·1 to 4·8 billion tons).[12] This is considerably less than the proved reserves of Saudi Arabia or almost half those of Kuwait and Iran. Except for these three countries, however, the CIA estimates that the USSR has larger reserves than anyone else in the world, including the United States.

Why are so few aware of the Soviet Union's role in the world petroleum market? To some extent Soviet importance is obscured by the fact that it diverts a little over 50 per cent of its exports to the members of the Council of Mutual Economic Assistance (CMEA), primarily those in Eastern Europe. And since close to 90 per cent of the oil imported by the CMEA countries (except for Rumania) comes from the Soviet Union, many specialists simply assume that CMEA is one large entity, at least as to its impact on the petroleum market. In addition, that April 1977 CIA report on Soviet petroleum has focused attention on the problems and petroleum imports the Soviet bloc will have in 1985. Accordingly, the CIA report has had the effect of diverting attention from the important strength enjoyed by the Soviet Union today.

What is lost sight of is that for the time being Soviet petroleum plays an important role in the world economy. With production as high as it is, Soviet petroleum is not only important within the Soviet Union itself and the CMEA but also in the outside world. As of the late 1970s, the Soviets were exporting about 23 million tons or not quite half a million barrels a day to the LDCs and to soft-currency countries like Finland and Portugal and, even more significant, about 1 million barrels a day or 50 million tons to the hard-currency countries. If all sales to the non-communist world are combined, Soviet export sales are comparable to those from countries such as Abu Dhabi, Algeria, Nigeria, Indonesia, or Mexico.

To state the situation in somewhat more dramatic terms, the Soviet Union is a one-crop economy, at least in terms of its exports to the hard-currency countries. As Table 1.1 indicates, in 1978 the Soviets derived $5·7 billion or over 50 per cent of their hard currency from petroleum sales. Nothing else in the civilian sector comes close to being that important. Some estimates indicate that military weapons may earn $2 to $3 billion a year, but outside of that the runners-up are natural gas and timber exports which each bring in only $1 billion. Everything else, including machinery exports, accounts for little more than a few hundred million dollars.

Since petroleum plays such an important role in Soviet foreign trade, it follows that the coming to power of OPEC has also had profound implications both internally and externally for the Soviet Union. Even though the Soviet Union is not a member of OPEC, what OPEC does is of immense consequence to it. More than that, because the Soviet Union exports so much petroleum, it is not an exaggeration to assert that it is one of the chief beneficiaries of OPEC. Thus, the price increases certified by OPEC in 1979 made possible enhanced hard-currency earnings for the Soviet Union of as much as $2·5 billion. But unlike Saudi Arabia, which serves as the chief regulator of OPEC's

Table 1.1 Exports of Major Soviet Commodities to the Hard-currency Countries in 1978 ($m.).*

	Austria	Belgium	Denmark	France	Great Britain	Greece	Italy	Japan	Netherlands	Spain	Sweden	Switzerland	USA	West Germany†	Other	Total	Finland
Coal	35		14	33			54	116	2	5	21			6		284	92
Oil and Oil Products	195	227	239	642	465	242	1,104	111	297	128	356	207	156	1,136	237	5,739	1,019
Nat. gas	221			140			254							480		1,094	68
Liqu. gas				11												11	
Chrome Ore				2							3		12	3		20	
Non-metallic																	
Mineral Alumina	3	44		14			2	6	3		9	23	2	23		128	9
Ferrous Metal	15		3		21	2	54	26	6		8	12		17	8	174	11
Chemicals	6	2	2	12	12	2	12	17	27		3		35	74	3	204	12
Potash	3	8			9		8	14			2		2		2	48	5
Timber and Timber Products	21	35	9	71	162	9	75	405	36	12	8		3	98	6	948	68
Cotton	11	5		122	27		6	125	6					33	3	336	9
Furs		2		62	62	2	6	11	5	3	3	11	8	20		123	2
Sunflower Oil	5			2	2				2				8	6		21	
Misc.		2	3	3	3		11	8						2	2	36	11
Products Total	513	336	257	1,044	762	255	1,584	836	384	153	411	252	225	1,893	260	9,164	1,301
Machinery	8	30	11	47	48	20	36	3	11	8	14	8	3	54	23	320	96
Export Total	557	612	290	1,260	1,286	284	1,668	1,104	450	170	468	429	380	2,247	287	11,489	1,506

* 1 ruble = $1·50.
† Includes West Berlin.

Source:
Ministerstvo vneshnei torgovli SSSR, *Vneshniaia torgovlia SSSR v 1978* (Moscow: Vneshtorgizdat, 1979). Hereafter *VT SSSR* with the appropriate year.

clout – which means Saudi Arabia often has to decrease or increase output, not because it wants to, but because it has to do so to make OPEC work – the Soviet Union derives benefits from OPEC at no cost or obligation to itself. In that sense it is reasonable to argue that it may even be *the* chief beneficiary of OPEC's actions.

Against this background, the release of the CIA report in April 1977 inevitably created a certain amount of confusion. If the Soviet Union is such a strong factor in the world energy scene, how is it so weak? And if the USSR does have an energy problem, why have Soviet authorities waited so long to come to grips with it? Why did the CIA release the report when it did? What is one to make of the so-called 'Swedish Report' which predicts almost the opposite of what the CIA did? According to the Swedes, by 1985, instead of the Soviet bloc becoming a net importer of 3·5 to 4·5 million barrels a day (MBD) the Soviet Union will be a net exporter to the hard-currency bloc of from 2·4 to 3·4 MBD.[13] What caused the CIA to qualify its original April 1977 report in July 1977 and to back down considerably in August 1979?

This much we do know, there were some unusual coincidences. The first time most of us heard that the CIA had prepared its far-reaching Soviet petroleum report was when it was mentioned in connection with the issuing by President Jimmy Carter of his first energy report to the Congress. The CIA was asked to project demand and supply conditions for petroleum in order to demonstrate the need for both increased conservation and increased forms of energy production, particularly substitutes for petroleum and gas. Going beyond typical energy studies which usually limit themselves to a survey of the market situation in the non-communist world, the CIA factored in the Soviet bloc as an active influence. As it was, market conditions in the non-communist world were serious enough. But when, from an already existing taut world market, the CIA subtracted yet another $1\frac{1}{2}$ million barrels a day of petroleum that the Soviets had been exporting, there was bound to be an even more dramatic effect. If the Soviet bloc not only finds itself forced to cut off $1\frac{1}{2}$ MBD of exports, but in addition finds it necessary to seek annual imports of 3·5 to 4·5 MBD, the Soviet action takes on crisis proportions. Such a switch would mean that the world petroleum market would somehow have to come up with an extra 5 to 6 MBD of petroleum on top of the demand increases anticipated from the non-communist world. Whether this 5–6 MBD is provided for by reducing the demands of existing non-communist consumers or by increasing the sources of supply, the point is that the shift is a major one for the market to absorb. For comparison's sake, remember that prior to its 1978–9 revolution Iran was exporting over 5 MBD, an amount equal to the potential of the Soviet impact.[14] The elimination of this petroleum from the world market set off the crisis of 1979. Even though the Soviet impact on the market would come in a more gradual manner, the ultimate disruption of the market could be equally far-reaching.

Whether intended or not, the CIA projection did heighten the urgency of the President's energy message. Unfortunately, however, the effect may have been counter-productive. For some, it did increase support for an energy program. But for a few, it may have served to reduce support. Broad sections of the public had long been skeptical about how genuine the energy shortage was. When some doubts were expressed regarding the validity of the CIA's conclusion about Soviet needs, this helped to undermine the credibility of the overall report. Furthermore, given the climate of the times, why did the CIA have to be brought in at all? The portion of the President's report referring to the USSR was less than a paragraph in length. For that matter, other government agencies could have made almost as good a case for an energy crisis without involving the Soviet Union or the CIA.

Some suspect that the CIA had a hidden agenda. Did the CIA intend to go beyond the world energy crisis and American energy policy to the realm of defense and political policy? Inevitably such a report on forthcoming petroleum shortages would make the Soviet Union seem much more vulnerable than it appeared to be in 1977. Such weaknesses would affect, not only Soviet internal industrial growth, but also its international, economic, and political posture. Indeed, some CIA studies, assuming a Soviet petroleum shortage, suggest exactly such conclusions.[15] On the basis of such assumptions, some commentators have argued that the United States should exercise stricter export controls in order to increase economic and political pressure on the Soviet Union.[16]

In the wake of the controversy created by the CIA report, while most observers focused on the Soviet Union's forthcoming problems, it often seemed than an equal number went on to assume that the Soviet Union was currently suffering from an energy shortage as well. Even careful readers who noted the CIA's prediction that the full impact of the Soviet shortfall would not hit until about 1985, eight years away, often managed to miss the fact that in 1977 the Soviet Union was the world's largest producer of petroleum, as well as the world's third largest exporter. Moreover, irrespective of what was to happen in the mid-1980s, the Soviet Union was likely to keep increasing its production of petroleum throughout the 1970s. It is true that at one point the rate of growth had fallen from 6 per cent or more to under 3 per cent, but what was lost sight of was that production was none the less increasing.[17] Whether or not the Soviet Union continues to increase its output or maintains production levels is clearly an important issue. One does not have to go to the extremes of the Swedish report. Yet if the USSR can just maintain production compared to the United States, it will find itself in an enviable position.

But what if the original CIA report proves to be correct? There is no doubt, as we shall see, that the Soviet Union is having production problems and trying to increase or even maintain its petroleum output. As reported by the CIA, new fields are not being found or exploited as fast as anticipated, yields from existing fields are falling faster than planned, and excessive waste pervades the system from the well to the engine. If these problems cannot be solved, then the implications for the Soviet Union, and for that matter for the rest of the world, will be very serious. Because it is so heavily dependent on the Soviet Union, the most serious effects will fall first and most severely on Eastern Europe. At the same time, there are likely to be consequences for several of the OPEC nations. Because the area is so volatile, it is hard to predict exactly what will happen, but the effect may be more political and military than economic in nature. Finally, because they are so dependent on imported energy and world energy developments, the OECD countries, including the United States, will also be severely affected.

The intent of this study is to examine the issues posed on the preceding pages. Is the Soviet Union faced with an impending petroleum shortage? If so, what are the internal and external consequences? Could it have a petroleum shortage and not have an energy shortage? What can it do to alleviate its shortages or simply to increase the output of energy? Given its importance, what explains why Soviet planners have been so slow to respond to the needs of the Soviet petroleum industry? What is the relationship between the anti-innovative Soviet planning system, the failure of Soviet planners to respond to the ever-growing problems in the petroleum industry, and the release of the CIA report? Can American export policy affect Soviet developments? What is likely to be the policy of our allies? Finally, what caused the CIA to make some fundamental revisions in its April 1977 report?

STRUCTURE

To provide the proper perspective it is necessary to go back in history in order to understand the role that petroleum has played in Russian economic development. In Chapter 2 we shall discover that petroleum production has a long history in Russia, and that this is not the first time that Russia has led the world in petroleum production. More than that, petroleum exports have played a vital role for over a century, even though it sometimes has meant dealing with questionable partners and relying on foreign technology and foreigners to perform tasks that Russian technology and the Russians themselves could not.

Aided by the discovery of a series of new and sometimes vast fields, we shall see in Chapter 3 that Soviet oil and gas production rose rapidly after the Second World War. To some extent that may have been unfortunate, since if reserves had not been so abundant Soviet officials might have devoted more effort to improving their development, production, and incentive systems. For many years there was every reason to believe that they could afford to be sloppy. Yet this carelessness may also have been fortunate. Inadvertently their careless methods may have spared the Soviets some greater problems in certain areas of energy policy. Western economists used to deride the Soviet price system for failing to facilitate the shift to liquid fuels from coal so long after the change took place in the West. Now it turns out that the Soviet Union may not have been so 'backward' after all.

In Chapters 4 and 5, we shall examine the Soviet Union's emerging role in the world energy market. Prior to 1973, most of the established oil producing states and companies regarded the Soviet Union as a market disruptor. The major exceptions were in Eastern Europe, where the Soviets were regarded rightly or wrongly as price exploiters of a captive market, and several LDCs, where the Soviets were viewed as friends for making it possible to obtain petroleum at a reasonable price in soft currency. By 1973, the Soviet Union's activities had evolved to such a degree that the USSR found itself uniquely positioned to take advantage of the 1973 embargo and the new world market price. With the newly expanded opportunities in the world market, the Soviet Union began to re-examine some of its export priorities. At the same time, however, its ability to find new reserves began to lag, and its production problems in existing fields began to mount.

Projecting these difficulties into the future, the CIA concluded that the Soviet Union would soon discover that its abundant petroleum had been depleted and that it had serious dilemmas to contend with both internally and internationally. We shall examine the CIA report in Chapter 6, and consider its validity, and see in Chapter 7 how the Soviet Union is coping with its energy needs. This will necessitate an examination of what the United States should or should not do to help the Soviet Union with its problems. We shall see that at the present time there is also a debate within the Soviet Union itself over just what Soviet policy should be, and whether or not any effort should be made to increase the production of all raw materials, not just petroleum.

Finally, in Chapter 8, we shall try to tie this together and ask why the CIA report appeared as it did, when it did. This will also give us a chance to consider the Swedish report and the equally mysterious circumstances surrounding its appearance. Most intriguing of all, we shall conclude with a look at the consequences of the appearance of the CIA report and the apparently unintended effect it has had in the Soviet Union, and how this in turn may have influenced the CIA itself to change its views.

For those who believe in the conspiracy theory of history, this study will probably prove to be disappointing. Unfortunately there does not appear to be any discernible conspiracy behind the release of the CIA report. Nor will we find CIA agents prowling around the Moscow vaults of the Ministry of the Petroleum Industry. Regrettably the story is not so dramatic. Yet unlike the normal economic analysis of either the Soviet economy or the world energy situation, the CIA's involvement in what otherwise might be a purely economic and engineering analysis provides an unexpected twist of controversy. What follows is the fascinating combination of oil, the Soviet Union, and the CIA.

NOTES: CHAPTER 1

1   *Economic Report of the President, 1979* (Washington: US Government Printing Office, 1979), pp. 183, 294.
2   Energy Perspectives, US Department of the Interior (Washington: US Government Printing Office, June 1976), pp. 63, 98, 190.
3   *New York Times*, 31 May 1979, p. D4.
4   Ray Vernon (ed.), *The Oil Crisis* (New York: Norton, 1976), pp. 15, 39.
5   Tsentral'noe statisticheskoe upravlenie, *Narodnoe khoziaistvo SSSR v 1977 godu* (Moscow: Gosstatizdat, 1978), pp. 62–5. (Hereafter *Nar. khoz.* and the appropriate year.)
6   CIA Handbook, 1978, p. 84.
7   loc. cit.
8   G. I. Martsinkevich, *Ispol'zovanie prirodnykh resursov i okhrana prirody* (Minsk: BGU, 1977), p. 64.
9   *Petroleum Economist*, April 1979, p. 177.
10 Central Intelligence Agency, *The International Energy Situation: Outlook to 1985*, ER77-10240 U (Washington: April 1977); Central Intelligence Agency, *Prospects for Soviet Oil Production*, ER 77-10270 (Washington: April 1977) (hereafter CIA Soviet, April 1977); Central Intelligence Agency, *Prospects for Soviet Oil Production: A Supplemental Analysis*, ER 77-10425 (Washington, DC: July 1977) (hereafter CIA Oil, July 1977).
11 *Oil and Gas Journal*, 26 December 1977.
12 CIA Oil, July 1977, p. 32.
13 PetroStudies Co., *Soviet Preparations for Major Boost of Oil Exports* (Malmo, Sweden: PetroStudies Co., 1978), pp. 45, 48.
14 *New York Times*, 9 March 1979, p. D6.
15 Central Intelligence Agency, *The Soviet Economy in 1976–77 and Outlook for 1978: A Research Paper* (Washington, DC: August 1978), ER 78-10512, p. 7 (hereafter referred to as CIA, August 1978.)
16 Carl Gershman, 'Selling them the rope. Business and the Soviets', *Commentary*, April 1979, p. 35.
17 Russian Research Center, Harvard University, Newsletter, 2 July 1979, Cambridge, Mass., p. 1.

2

History of Energy and Petroleum in Russia

There is something about petroleum that is controversial and intriguing. There is something about Russia that is mystifying and absorbing. When the two merge in a study of Russian petroleum, the result is likely to be tantalizing and engrossing. In the same spirit this historical preface, which will carry us up to 1945, is more than a perfunctory filler to the main analysis. There are so many precedents, similarities, and coincidences in a study of the history of Russian petroleum that discussion of the present generates a sense of *déjà vu*.

THE EARLY YEARS

Although they were unaware of its ultimate potential at the time, seventeenth- and eighteenth-century residents of what was to become Baku knew about and used the region's petroleum and natural gas. This is indicated by the appearance in the region centuries ago of the Parsees, who worshipped fire.[1] These followers of Zoroaster had built a temple 7 miles outside Baku which served as a holy site until 1880. Its perpetual flames were probably fed by natural gases escaping from the abundant deposits under the temple.[2] Even Marco Polo during his thirteenth-century travels noted that traders were very active in carrying oil-soaked sand to Baghdad.

Central Russian influence on Baku appeared relatively late. After the fall of Constantinople, control of the Black Sea fell to the Turks, who kept the Russians out of the area for several centuries. On the other side of the Caucasus the Persians had control of the Caspian Sea. Ivan the Terrible pushed Moskovy's influence down to Astrakhan on the north shore of the Caspian Sea in the sixteenth century, but formal Russian control of Baku did not come until the conquest of the area by Peter the Great in 1723. Once in command, Peter sought to ship some of the region's oil to St Petersburg for possible use, but his advisors thought it was not worth the effort.[3] It did not matter much since shortly thereafter, in 1735, the Persians regained Baku and impeded what little petroleum trade with the north there was. It was only in 1806 that the Russians recaptured Baku and in 1813 that they finally signed a peace treaty.[4]

Before the arrival of the Russians petroleum extraction was very primitive. For centuries petroleum traders had to extract the petroleum with rags and buckets. As some of the pits increased in depth, a hand-winch was occasionally used, but otherwise the work was quite unsophisticated. The Russians were able to improve the technology somewhat and production increased accordingly. In 1848 a Russian, A. F. Semenov, drilled the first relatively deep well. But even then the well was only 20 to 30 yards deep.[5]

In 1821, after their reconquest of the area, the Russians set up a special franchising system for those who wanted to produce and sell petroleum. The rights to drill and pump petroleum on a specific site were extended on a monopoly basis for four-year periods.[6] However, the lease could be revoked at the end of that time and there were no options for renewal. This precluded most serious exploration and drilling activity and caused the leaseholders to extract as much as they could during the four years of their lease with little or no thought about maximizing the long-run output of the area. This system prevailed until January 1873, when a more efficient public auction system was introduced.[7] As Table 2.1 indicates, the changes facilitated a sharp increase in production. Although it was small to begin with, production doubled the following year. The discovery of Baku's first gusher in the early 1870s facilitated this growth.[8]

Table 2.1   *Early Russian and American Oil Production and Exports*
(thousand metric tons)

	Production	Export	US Production
1860	4		
1861	4		
1862	4		
1863	6		
1864	9		
1865	9	2	340
1866	13	2	
1867	17	3	
1868	29	2	
1869	42	1	
1870	33	2	715
1871	26	1	
1872	27	2	
1873	68	1	

1874	106	2	
1875	153	2	
1876	213	2	1,242
1877	276	1	
1878	358	1	
1879	431	5	
1880	382	3	3,575
1881	701	18	3,762
1882	870	19	4,128
1883	1,039	59	3,189
1884	1,533	113	3,294
1885	1,966	178	2,973
1886	1,936	247	3,817
1887	2,405	311	3,846
1888	3,074	573	3,755
1889	3,349	734	4,782
1890	3,864	788	6,232
1891	4,610	889	7,384
1892	4,775	937	6,870
1893	5,620	985	6,587
1894	5,040	880	6,710
1895	6,935	1,059	7,193
1896	7,115	1,058	8,290
1897	7,566	1,046	8,225
1898	8,635	1,115	7,530
1899	9,264	1,392	7,762
1900	10,684	1,442	8,652
1901	11,987	1,559	9,468
1902	11,621	1,535	12,072
1903	11,099	1,784	13,662
1904	11,665	1,837	15,923
1905	8,310	945	18,322
1906	8,885	661	17,203
1907	9,760	733	22,589
1908	10,388	797	24,280
1909	11,248	796	24,911
1910	11,283	859	28,500
1911	10,547	855	29,981
1912	10,408	839	30,319
1913	10,281	948	36,144
1914	10,013	529	38,230
1915	10,138	78	40,904

THE NOBEL BROTHERS

These developments in turn attracted other prospectors, particularly foreigners like the Norwegian Robert Nobel. Nobel in particular came to exercise enormous influence in the area not only as a producer but as a refiner and marketer.[9] By 1883 production exceeded 1 million tons; by 1887 it exceeded 2 million tons. Equally significant, in 1877 and again in 1882, 1885, and 1891, strong and increasingly effective tariffs were passed which helped to curb Russian imports of American kerosene.[10]

However, it took more than tariffs to stem the flow. Standard Oil of the United States, as the world's largest producer, had staked out a dominant share of the Russian market. Before Standard Oil could be pushed out, some way had to be found to facilitate the shipment of petroleum and kerosene from Baku to the urban centers of Moscow and St Petersburg. Just as during the Crimean War it was easier to move troops from Paris and London to the Crimea than from St Petersburg and Moscow, so it was easier to move kerosene from the United States to St Petersburg than it was from Baku. One of the Nobel brothers' most innovative accomplishments was to facilitate the flow of Russian oil north. To do this Robert's brother Ludwig designed a pipeline to carry the crude oil from the well to their refinery and then to the Caspian Sea. In 1878, to carry large enough quantities to the Caspian Sea and to make the venture profitable, he also conceived of and constructed the first oil tanker, the *Zoroaster*.[11] His tankers docked at Astrakhan, where the oil was transferred to barges which then moved up the Volga. A storage depot was established in Tsaritsyn (later to become Stalingrad and now Volgograd) where by 1881 it became possible to reload the petroleum on railroad cars, a convenience that was particularly important in the winter when the Upper Volga was frozen. The end result of Nobel's innovation and the government's higher tariffs was the all but complete halt of kerosene imports from the United States. Imports which were 4,400 tons in 1884 fell to 1,130 tons in 1885 and to an almost unnoticeable 22 tons in 1896.[12]

The cultivation of domestic markets was followed by an effort to expand foreign markets. For obvious geographical reasons Persia had always been a major consumer of Baku's oil. For equally obvious geographical reasons it was difficult to supply other regions of the world, including St Petersburg. One of the first challenges was to break through the barrier of the Caucasus Mountains in order to gain access to the Black and Mediterranean Seas and thus to the ocean routes beyond.

It was only in 1878 when the Russians pushed the Turks out of Batum on the Black Sea that a new route became a possibility. Shortly thereafter a group of Russian oil producers in the Baku region, led by Bunge and Palashkovsky, obtained a franchise to build a railroad over the mountains from Baku through Tbilisi to Batum on the Black Sea. Since they were short of funds, they sought the help of the Nobels. Shortsightedly, the Nobels refused, fearing that their dominance of the

Baku trade, especially their St Petersburg markets, would be jeopardized by the additional competition. Not to be denied, Bunge and Palashkovsky turned instead to the French house of Rothschild. Having recently backed a refinery on the Adriatic, the Rothschilds were anxiously searching for a source of crude oil to free themselves from dependence on Rockefeller's Standard Oil.[13] The cork on Russian exports was pulled when the railroad was completed in 1883–4. Table 2.1 indicates how overall exports increased. Exports from Batum, which totalled 3,300 tons in 1882, increased to 24,500 tons in 1883 and 65,000 tons in 1884, an amount equal to previous total exports from all Russian ports.[14] The flow soon became even greater once a 42-mile pipeline replaced the most rugged portion of the railroad route in 1889.[15]

RUSSIA AS AN OIL EXPORTER

The increased flow of Rothschild's petroleum from Batum and Nobel oil via the Volga meant competitive pressure on Standard Oil's markets in England. Angry over the threat to his English and European markets by the intruding Russian oil, Rockefeller and Standard Oil retaliated in what was to become a familiar pattern by cutting prices. For a time this tactic succeeded but ultimately the Russian producers prevailed and carved out a share of the market for themselves. Whereas the combined Rothschild–Nobel share of the British market amounted to only 2 per cent in 1884, by 1888 it had expanded to 30 per cent.[16] Overall, however, compared to American exports, Russian exports were relatively more important only in Asia. Thus in 1897 75 per cent of American exports went to Europe and 16 per cent to Asia, whereas only 59 per cent of Russian exports went to Europe but 38 per cent went to Asia.[17] The pattern was much the same in 1913. (See Table 2.2.)

For a long time the petroleum flowed so readily in the Baku region that there seemed to be no reason to exercise much care in exploring for new fields and pumping existing fields. The waste was enormous, not to mention hazardous. American visitors in what over the years was to become a standard theme were amazed by the inefficiency, sloppiness and lack of care of Russian petroleum operators.[18]

Still, there seemed to be little need to worry as long as the oil kept flowing. Moreover, as is also true in the 1970s, the per capita consumption of oil, or more appropriately petroleum products, was much lower in Russia than it was in other advanced countries in the world. This was due in large part to the lower standard of living in Russia. In the late nineteenth century, for instance, Russian consumption of kerosene was one-half of that in Germany.[19] And since

Table 2.2   *Russian Petroleum Exports in 1913 (tons)*

Afghanistan	230
Austria	3,938
Belgium-Luxemburg	83,781
Bulgaria	3,725
China	3,373
Denmark	8,210
Egypt	123,263
Finland	50,995
France	112,591
Germany	129,429
Great Britain	3,725
Greece	3
Iran	34,813
Italy	16,577
Mongolia	2
The Netherlands	33,363
Norway	640
Rumania	4,639
Spain	2,675
Sweden	5,118
Turkey	149,024
United States	245
Yugoslavia	6,581

domestic productive capacity exceeded domestic petroleum needs, petroleum producers generally sought to divert a portion of their output to foreign markets. For example, during the good production years of 1903 and 1904, the Russian-based producers exported 16 per cent of their total production. (See Table 2.1.) It was in 1904 that absolute petroleum exports reached their peak of 1·8 million tons. However the relative share of petroleum exports had been even higher in 1890 and 1892, when 22 per cent of all petroleum produced was exported.

Not surprisingly, therefore, Russian petroleum exports often exceeded those of the United States during the late 1890s and the early twentieth century. And if they were not the largest exporter, the Russians were certainly the second largest. It is difficult to tell precisely which years the Russians out-exported the Americans because the data are incomplete. While export–import data for crude oil are available for comparison, the United States statistics on imports and exports of petroleum products begin only in 1940.[20] While net American exports of refined products were high, American imports of crude oil during the 1920s were generally even higher.

In any case, as Table 2.1 indicates, Russian production exceeded

American production from 1898 to 1902. During this period Russia was the largest producer of petroleum in the world. At the time the Middle East was only a desert. It was 1938 before oil was discovered in Saudi Arabia. The only other oil producing areas of note at the turn of the century were in the Netherlands, East Indies, and Mexico. Even with that, in 1897 Russia and the United States accounted for about 95 per cent of the world's production.[21]

PRODUCTION STAGNATION

The high point for Russia was in 1901 when Russian production reached a pre-revolutionary peak of 11·987 million tons. The comparable figure for the United States was 9·468 million tons. (See Table 2.1.) But while American production of crude oil reached 12 million tons the following year and continued to climb every year but one until 1924, it took Russian production until 1929 to exceed the 1901 level. What explains the decline in Russian production which took place after 1901 ?

Initially there were rumors that the fields of Baku were running dry. It appeared that these rumors were well founded since, as we saw, production practices were very wasteful.[22] But prior to the Second World War at least, the fall in output was not due solely to the exhaustion of the Baku fields. None the less, the rumors about Russian production problems were not easily dispelled. As a forerunner of what was to come, the Russian scientist Dmitry Mendeleyev wrote a paper entitled 'The supposed exhaustion of the Baku oilfields'.[23]

The Russians sought to cope with the drop in output in existing wells in a variety of ways. First, as production fell in some of the older Baku fields, new fields were drilled nearby. Secondly, new deposits outside the Baku region were discovered. Whereas Baku accounted for 96 per cent of all Russian production in 1897, by 1910 it made up 85 per cent and by 1913 even less.[24] Most of the difference was accounted for by the new fields opened up at Grozny, 300 miles to the north at Emba on the northern shore of the Caspian Sea, and at Maikop, only 50 miles from the Black Sea. Although there is some reason to believe that the existence and even early production of some of these sites as at Grozny predated the arrival of the West Europeans, many of the more important fields were developed by foreigners, especially with English capital.[25] Technology also was improved in part with help brought in from the West. Learning how to drill deeper produced the quickest results. The commonly used Russian drilling methods, which often relied on wooden, not metal, tools, made it difficult to go deeper than 300 feet. By the end of the nineteenth century, Nobel and some of the other foreign concerns were drilling wells over 600 yards deep. With the help of the American-produced rotary drilling system, by 1909 the wells were reaching 800 yards.[26]

Yet ultimately the Russians could not prevent a sharp decline in their production and exports. In part the drop was due to failure of the Russian companies to import enough technology. Refining and drilling techniques began to change rapidly in the early twentieth century and few of the Russian companies kept up. The Russian oil companies also were swept up into the international corporate scheming and rivalry of the period. Inevitably the jockeying for a market share by companies such as Standard Oil and Shell had some impact. Price-cutting was a common tactic in increasing control. As a result many producers cut back on their operations and occasionally went bankrupt. Recurring depressions had the same effect. Nor did all the manipulating take place outside Russia. In 1911 Shell became a major factor in the Russian market when it purchased the Rothschild family's petroleum holdings. Since the revolution and expropriation were only six years away, it was probably one of the smartest sales the Rothschilds ever made. But most deal-making of this sort involved more financial than technological innovation and thus added little to the country's productive capabilities.

Also hurtful was the decision by the Czarist government in 1896 to change the concession system that had governed Russian oil production.[27] In hopes that it could collect more revenue, the government instituted a combined auction royalty system. This seems to resemble the system that was ultimately widely adopted in the Middle East in the 1950s and the 1960s, although the royalties demanded by the Russians sometimes reached as high as 40 per cent. Aided by the wisdom accumulated over eighty years, the Russian system of those days would look like a bargain today for a foreign investor. But considering that until seventy-five years later no one else imposed what then seemed to be such confiscatory terms, the change was opposed by the concession-holders and they reduced their output.

More damaging than anything else, however, was the growing labor and civil unrest which hit the Batum and Baku areas. Led in part by Stalin, strikes occurred in the Batum area as early as 1901–2.[28] That was followed in July 1903 by an oil worker strike in Baku. Interspersed between almost annual strikes in 1904, 1905 and 1907 were the activities of the reactionary Black Hundreds. What the strikers did not pillage or burn, the Black Hundreds did. Nor did the complex racial mix of the Tatars, Armenians, Jews, Russians, and Muslim Turks and Persians in the area add to the tranquility of the region once tensions erupted. The climax came during the 1905 Revolution when overall production fell over 3 million tons and exports were cut in half. Neither production or exports were to recover in any significant way until long after the 1917 Revolution.[29]

The rather disappointing years of production and export in the last decade before the revolution should not obscure the fact that the petroleum industry in pre-revolutionary Russia had an important role to play. (See Table 2.1.) Not only did Russia produce more petroleum than any other country for a short period of time, but there were periods when petroleum contributed in a fairly important way to the country's export earnings. True, petroleum exports never came close to matching grain export earnings which accounted for 50 to 70 per cent of the country's export earnings from 1895 to 1914.[30] But except for timber, petroleum was often the second non-agricultural export. In 1900 and 1901, petroleum generated 7 per cent of all the country's export earnings. That was a high point, particularly after labor troubles disrupted production and exports, but for a time at least the large exports were a foretaste of the role petroleum was to play after the revolution as well.

THE REVOLUTION

The 1917 Bolshevik Revolution had an immediate impact on oil production. Output fell from 10·8 million tons in 1916 to 8·8 million tons in 1917. This reflected the unrest caused by the workers' insistence on more control over managerial decision-making. In some cases workers' committees were formed to superintend management. Naturally this led to considerable confusion. Formal confiscation was declared the following year, on 6 June 1918, when the fields were officially nationalized.[31] Then production fell to 4·1 million tons. (See Table 2.3.)

The path of recovery was a difficult one and it involved the opposition as well as the support of various foreign companies. We saw that one of the more notable aspects of the pre-revolutionary period of Russian oil development was the important role played by foreigners. Swedish, French, British, and even American investors and operators had devoted large sums of money in an effort to gain control and increase production. With the exception of the Rothschilds, the revolution meant a loss for most of them. The decade that followed was marked by the effort and intrigue of many of the former foreign operators to out-maneuver the newly empowered Bolshevik rulers in order to regain or repatriate some of their money. Even with the help of foreign military intervention most failed, but oil men have always been more venturesome than most of us.

The Turkish occupation of Baku in September 1918 provided the opening the old investors had been waiting for. Aware that the Bolsheviks were very much distracted by unrest in the north, the British sent in an expeditionary force and in November 1918 pushed out the Turks. The

Table 2.3   *Petroleum Exports and Production, 1913–41 (tons)*
EXPORTS

	Crude	Refined	Total	Production*
1913	739	951,811	952,550	10,281
1918	39	1,677	1,716	4,146
1919	—	13	13	4,448
1920	—	85	85	3,851
1921	3	7,391	7,394	3,781
1921/2	1,491	50,763	52,254	4,658
1922/3	35,431	48,863	84,294	5,277
1923/4	74,441	637,223	711,664	6,064
1924/5	63,471	1,309,024	1,372,495	7,061
1925/6	116,759	1,356,728	1,473,487	8,318
1926/7	161,276	1,924,868	2,086,144	10,285
1927/8	248,466	2,534,375	2,782,841	11,625
Oct.–Dec. 1928	42,541	753,360	795,901	
1929	310,264	3,548,329	3,858,593	13,684
1930	293,659	4,419,374	4,713,033	18,451
1931	382,175	4,843,276	5,225,451	22,392
1932	525,915	5,591,749	6,117,664	21,414
1933	525,518	4,404,609	4,930,127	21,489
1934	458,519	3,856,712	4,315,231	24,218
1935	206,688	3,161,931	3,368,619	25,218
1936	166,555	2,499,041	2,665,596	27,427
1937	68,553	1,860,778	1,929,331	28,501
1938	167,794	1,220,565	1,388,359	30,186
1939	244	473,977	434,221	30,259
1940	—	874,301	874,301	31,121
1941				

* Production figures in thousand tons and for calendar years 1922, 1923, 1924, 1925, 1926, 1927, 1928.

British apparently had hopes of setting up an independent state of Azerbaidzhan. This was not solely an anti-Bolshevik gesture, but also an anti-Russian step to protect Persia and block Russian access to British India. Much the same type of maneuvering took place after the Second World War, only this time the Soviets were attempting to do the extending from Azerbaidzhan into the northern part of Persia/Iran.

With the British takeover and denationalization of the oil fields, hopes in the European stock markets soared on the expectation that the weak Bolsheviks would never come back. Moving fast in hopes that it could establish a presence in the area where previously it had been weak, Standard Oil of New Jersey signed a contract in January 1919 with the independent government of Azerbaidzhan.[32] It paid one-third

Petroleum Exports and Production, 1946–78 (million tons)
EXPORTS

	Crude	Refined	Total	Production
1946	—	0·5	0·5	21·7
1947	—	0·8	0·8	26·0
1948	—	0·7	0·7	29·2
1949	0·1	0·8	0·9	33·4
1950	0·3	0·8	1·1	37·9
1951	0·9	1·6	2·5	42·3
1952	1·3	1·8	3·1	47·3
1953	1·5	2·7	4·2	52·8
1954	2·1	4·4	6·5	59·3
1955	2·9	5·1	8·0	70·8
1956	3·9	6·2	10·1	83·8
1957	5·9	7·8	13·7	98·3
1958	9·1	9·0	18·1	113·2
1959	12·5	12·9	25·4	129·6
1960	17·8	15·4	33·2	147·9
1961	23·4	17·8	41·2	166·1
1962	26·3	19·1	45·4	186·2
1963	30·2	21·1	51·3	206·1
1964	36·7	19·9	56·6	223·6
1965	43·4	21·0	64·4	242·9
1966	50·3	23·3	73·6	265·1
1967	54·1	24·9	79·0	288·1
1968	59·2	27·0	86·2	309·2
1969	63·9	26·9	90·8	328·4
1970	66·8	29·0	95·8	353·0
1971	74·8	30·3	105·1	377·1
1972	76·2	30·8	107·0	400·4
1973	85·3	33·0	118·3	429·0
1974	80·6	35·6	116·2	458·9
1975	93·1	37·3	130·4	490·8
1976	110·8	37·7	148·5	519·7
1977			155–9	545·8
1978				572·0

of a million dollars for drilling sites. The Nobels toyed with the idea of selling their shares to the Anglo-Persian Oil Company, but quickly grabbed yet another offer from Standard Oil. A tentative agreement was signed on 12 April 1920. Despite the fact that the Bolsheviks retook the area later that month, Standard remained convinced the Bolsheviks would not be able to hold on. Reflecting its confidence, it paid Nobel half a million dollars for some additional land. Ultimately Standard Oil paid Nobel several million dollars for its stock which, of course, had already become worthless. According to Tolf, this Standard Oil purchase was later to constitute one-tenth of the entire American claim against the Bolsheviks.[33] However, this speculative fever was not limited to Standard Oil. Shell Oil, along with other European investors, also bought what turned out to be worthless shares.

While the Soviets had gained physical control over the territory, they soon discovered that without the technical and managerial help of foreigners and others who had fled the area they could not really operate the oil fields. Output continued to fall until it reached a low in 1921 of 3·781 million tons, a level not seen since 1889. To add to their headaches, the Bolsheviks also found that the Western oil companies had united to boycott Russian oil exports, a pattern that was fairly common whenever oil fields were nationalized and often successful, at least until the late 1960s.[34] Formed in mid-1922, the Front Uni represented an oil consortium of fifteen companies. But through Western greed and Russian negotiating skill, the Front Uni's embargo was broken even before it began to operate. Shell, itself a leader of the boycott, made a purchase of Russian oil in February 1923 and the French followed soon after.[35]

THE FOREIGNERS RETURN

The crack in the embargo had appeared even earlier. The Soviets began to solicit foreign production help and the oil companies responded. Lenin personally approved such measures under the New Economic Policy (NEP) which authorized extending concessions for foreigners. One of the first to respond to the Soviet request for help was an American company, the Barnsdall Corporation.[36] Signed in October 1921, the Barnsdall contract actually predates the embargo. This was an important breakthrough for the Soviet Union. Not only did Barnsdall help the USSR restore production, it served to attract several other foreign companies, including British Petroleum, the Societa Minerere Italo Belge di Georgia, and eventually a Japanese group in Sakhalin.[37] Once the breach had been made the embargo was destined to fail.

The foreigners did what they were supposed to do. Oil fields were restored and new ones started. Barnsdall brought in advanced rotary drills and deep well pumps. Production rapidly recovered and, although there is some uncertainty as to how much Barnsdall made out of the venture, by 1924, when it left the Soviet Union, production was back up to 7 million tons. (See Table 2.3.) Production continued to increase, as did foreign technical help. Besides work at the wells, foreign help included American, German, and British assistance in the building of a second pipeline from Baku to Batum, the French supply of a

Schlumberger well-logging process and American (Standard Oil of New York), German, and British support for refinery construction.[38]

Once output had recovered the Soviets began systematically revoking their concessions. By December 1930 most of them had been closed out.[39] Standard Oil, however, was allowed to retain its concession at the kerosene refinery built in Batum until at least 1935 and the Japanese stayed on Sakhalin until 1944.[40]

Soviet policy was not consistent. Almost at the time the Soviets were closing down Standard's Oil's concession, they issued a new series of contracts. Sutton records how companies like Badger, Universal Oil Products, and Lummus were called back to rebuild and reconstruct refineries.[41] Some of their work continued until 1945, when it was supplemented with Lend-Lease contracts. With all this help, production and exports rose rapidly. Output increased by almost 5 million tons in 1929 and by about 4 million in 1930, so that production reached the level of 22 million tons, almost double the pre-revolutionary peak. (See Tables 2.1 and 2.3.)

As production increased so did the amount of administrative control emanating from Moscow. In the early days, however, there was more control in principle than in practice. Theoretically control over industry was centered in the Supreme Council of the National Economy (VSNKH), which was set up shortly after the revolution in 1917. VSNKH in turn derived its power from the Council of People's Commissars (CPK).[42] The CPK (the forerunner of the Council of Ministers) in turn created the Chief Oil Committee (*Glavny Neftianoi Komitet*) under the VSNKH on 17 May 1918. This was a grand meaningless gesture since it was only a few months later that the Turks and then the British pushed the Bolsheviks out.

Meaningful control had to wait until the British and Turks were sent home in the spring of 1920. Then, recognizing the communication problem between Moscow and the Caucasus, the Chief Oil Committee authorized the creation of three local operating trusts. Azneft, which apparently was the most efficient and aggressive, took over control of the Baku region, Grozneft took over Grozny, and Embaneft took over the fields in the Emba area.[43] The three trusts in 1922 formed a commercial syndicate, Neftesyndikat (later succeeded by Soiuzneft) to handle exports and other foreign activities.[44] Neftesyndikat proved to be a very aggressive monopoly. It joined together in a fifty-fifty partnership in 1923 with the English firm Sale & Company to market oil in the United Kingdom.[45] Neftesyndikat reserved the right to buy out all Sale Company shares in ten years. This first British company was followed by the second. This time the partnership was between Neftesyndikat and Royal Dutch Shell. The Soviets also entered an arrangement with Standard Oil of New York to market Russian oil in the Near and Far

East. Other deals were made with British-Mexican Petroleum, Asiatic Petroleum, and Bell Petrole.

Neftesyndikat kept expanding and set up the Russian Oil Products (ROP) company in London jointly with Arkos. By 1925 Russian Oil Products had its own filling station network. The Soviets also set up a filling station network in Germany called Derop through its subsidiary Deutsche-Russische Naptha Company. Ultimately other wholesale and filling station subsidiaries were formed in Sweden, Spain, Portugal, and Persia. With such a network to supply, Soviet oil exports increased rapidly. As Table 2.3 shows, Soviet oil exports surpassed the previous level in 1926–7 even though the production level was not exceeded until two years later. Reclaiming and in some cases going beyond their pre-revolutionary penetration, Soviet oil had an important impact in world markets. According to W. Gurov, who at the time was chairman of Soiuznefteeksport, the successor organization to Neftesyndikat-Soiuzneft, Soviet oil exports accounted for 14 per cent of all West European imports from 1926 to 1935.[46] The Soviet share reached a high of 17 per cent during the USSR's peak export years of 1929 to 1933 when the sum of exports amounted to 24·8 million tons. (See Table 2.3.) Soviet statistics also show sales to the United States of as much as 50,000 tons in the peak year 1930. Of course, that might not have gone to the United States directly, but to Standard Oil for reshipment to the Middle and Far East.

The most important purchaser in terms of physical volume and market share was Italy. According to Gurov's calculations, Soviet oil accounted for 48 per cent of Italy's total oil imports during the ten-year period from 1925 to 1935. The economic significance is clear, but it also was of political importance because as of 1922 Mussolini had become Prime Minister and in 1925 he became dictator. In other words, politics, at least in this instance, was no barrier to export. Only in 1938 and 1940 did the Soviet Union not export petroleum to Mussolini's fascist government.[47] The Soviets were only slightly more discreet in selling petroleum to Hitler's Germany. Sales remained at the relatively constant level of 400,000–500,000 tons until 1936.[48] Exports then fell to about 350,000 tons in 1936 and to 275,000 tons in 1937. In 1938 and 1939 they dropped to almost nothing but shot back up to 657,000 tons in 1940 after the signing of the Nazi–Soviet Pact. In fact, in 1940 Soviet sales to Nazi Germany accounted for 75 per cent of all Soviet petroleum exports that year.

PETROLEUM AND THE SOVIET BALANCE OF TRADE

Given their magnitude, Soviet petroleum exports were important not only for the purchaser but for the Soviet balance of trade. Whereas before the revolution petroleum exports at the highest accounted for 7 per cent of Russia's export earnings, in 1932 Soviet petroleum earnings generated 18 per cent of total Soviet export receipts. Exports in that record year amounted to 6·1 million tons and accounted for 29 per cent of total production. Soviet net exports of petroleum far exceeded American net exports in 1932–3. It should be pointed out that while the 18 per cent share of exports in part was due to the increase in petroleum exports, it was also due to the sharp fall in grain exports. Remember that at one time in the nineteenth century grain accounted for 70 per cent of Russia's export earnings. Russia was truly the bread basket of Europe. However, by the twentieth century, Russia's role as a grain exporter had diminished so that grain brought in only 53 per cent of the country's export earnings. None the less, with the advent of communism and especially collectivization, grain never again ever amounted to as much as 22 per cent of the export volume and usually amounted to about 10 per cent.

Important as petroleum was, however, it was not as important as some observers have thought it was. Sutton asserts that petroleum exports 'became a significant factor in Soviet economic recovery, generating about 20 per cent of all exports by value; the largest single source of foreign exchange.'[49] Later on he refers specifically to 1928 when he says petroleum amounted to 19·1 per cent of all earnings.[50] His assertions are then repeated by Tolf and Segal.[51] However, Sutton exaggerates somewhat. In 1928 petroleum accounted for about 14 per cent of all earnings.[52] Moreover, the relative earnings of timber exports exceeded those of petroleum throughout the 1920s and 1930s, often by a substantial margin. For that matter there were years such as 1922–3, 1926–7, 1930, 1931, 1937, 1938 and 1940 when grain earned more than petroleum.

The fall-off in petroleum exports after 1932 was a part of the general reduction in all exports that began at that time. From a peak of 3·2 billion rubles in 1930, Soviet export revenues fell to 2·8 billion rubles in 1931 and kept falling yearly (except for 1937), until they reached a mere 462 million rubles in 1939. (All trade figures cited are stated in terms of constant 1950 ruble prices.) They rose briefly in 1940, but again this was a by-product of the Nazi–Soviet Pact. Exports to Germany amounted to 50 per cent of all Soviet exports that year. In part, some of the reason for the drop in exports was that the Soviet Union began to need more of its raw materials for its own domestic production needs. More important, Soviet efforts were undercut by the Depression. Depression may be a capitalist disease and it may have no ostensible effect on the internal workings of the Soviet economy, but there is no denying that it has a devastating effect on the world demand for raw materials. This in turn affects the prices. The Soviets quickly realized that they were exporting more, but earning less. Thus 4·7 million tons of petroleum exports earned them 548 million rubles in 1930, while 6·1 million tons earned them only 375 million rubles two years later. Virtually all Soviet exports were affected in the same way.

ORGANIZATIONAL CHANGES

The early 1930s also marked a series of changes in the organization of the petroleum industry as regards both its foreign trade and producing functions. Having been joined together as the Commissariat of Foreign and Domestic Trade in 1925, the Commissariat of Foreign Trade was broken out as a separate unit in 1930. Shortly thereafter, on 26 February 1931, Soiuznefteexport, the successor to what originally was Neftesyndikat, was formally constituted as a foreign trade organization under the Commissariat of Foreign Trade.[53] Except for another short-lived merger of the Ministries of Domestic and Foreign Trade in 1953 and 1954, there have been only a few cosmetic changes since. (See Appendix 1.)

In contrast, there have been a host of reorganizations affecting oil production and refining. The Council of the National Economy (VSNKH) was abolished on 5 January 1932. Three people's commissariats replaced it: one for the timber industry, one for light industry, and one for heavy industry. The Committee on Oil was attached to the People's Commissariat of Heavy Industry. The next change occurred on 2 January 1939, when the Commissariat of Heavy Industry was broken into several parts, including a Commissariat of the Fuel Industry.[54] This lasted ten months, until 12 October 1939, when the Fuel Commissariat was divided into the People's Commissariat of the Coal Industry and the People's Commissariat of the Petroleum Industry.

After the war, as Stalin attempted to present a less threatening image, all people's commissariats were abolished and on 15 March 1946 turned into ministries. A few weeks later, on 4 April 1946, the Ministry of the Petroleum Industry was split into the Ministry of the Petroleum Industry of the South and West Regions and the Ministry of the Petroleum Industry of the Eastern Region, which focused primarily on Sakhalin. This also lasted a short time. On 28 December 1948 the two parts were reconstituted, and the Ministry of the Petroleum Industry was left intact until Stalin's death and Khrushchev's massive abolition of almost all ministries.

In a burst of reorganizational activity in 1957 Khrushchev did away with central control and instead created 110 regional *sovnarkhozy*. An effort was made to transfer all power from Moscow. The ministries were abolished and their functions decentralized and spun off to the *sovnarkhozy*, although obviously petroleum extraction was only important in a limited number of them. Even though the emphasis was on decentralization, some of the central functions were none the less held on to by Gosplan (the State Planning Organization) and various state production committees that evolved over time and were subordinated to Gosplan. Khrushchev, however, kept changing Gosplan's own functions as well as the number of *sovnarkhozy* and their functions. As a result, industrial and planning organizations were often in a state of flux. One of the charges against Khrushchev, when he was removed from office in 1964, was that 'his harebrained ideas' had caused enormous administrative confusion. Ultimately, in 1965, the industrial ministry system was restored by Khrushchev's successors. This involved untangling the several state production committees that had sprung up in the wake of the industrial ministry's abolition. As of 1965, there were state production committees for the fuel industry, for the gas industry, for petroleum extraction, and petroleum refining. These had all spawned from the Committee for the Chemical and Petroleum Industries. These four committees were ultimately restored into six ministries; the coal industry, the gas industry, the petroleum extraction industry, the petroleum refining industry, the petrochemical industry, and the chemical industry. Except for the renaming of the Ministry of the Petroleum Extraction Industry as simply the Ministry of the Petroleum Industry in 1970, that system continues to the present day.

If this seems like a roundabout route to end up with the same simple Ministry of the Petroleum Industry which existed prior to 1957, it is necessary to remember that most of the reorganizations were more a product of the obligatory cycle of bureaucratic juggling that Soviet leaders resort to than any problem specific to the petroleum industry. The only substantial change involved the creation of a separate entity for the gas industry in 1956 and the breaking-off of the petroleum refining and petrochemical ministries. Otherwise most of the other changes were due to the counterposing of centralization and decentralization. It is like the industrial consultant at the Harvard Business School who, when called upon to prescribe a cure for an ailing company, asks for the existing table of organization. When he determines that the company is centralized, he recommends decentralization. This is not only sound medicine, but it assures the consultant of an income for some time to come. He knows that before too long he will be called in again because new problems are bound to develop. Next time, however, his solution will be to centralize.

Yet there were some problems peculiar to the oil industry which in part had nothing to do with the organization of the People's

Commissariat of the Petroleum Industry. These problems were more a reflection of the fact that by the end of the 1930s the yield from the Baku fields had begun to diminish. Production fell temporarily in 1932, rose again impressively until 1938, but then leveled off. As a result petroleum output lagged far behind the targets set out for the Second Five-Year Plan, which ended in 1938. Production totaled 30 million tons, significantly behind the 46·8 million goal.[55] With the technology then at their disposal, the Soviets could not increase the production rate at their traditional fields in Baku and at Grozny.

Yet just as production seemed to be tapering off in the Caucasus, important new fields were discovered in the region between the Volga and the Urals. Called the 'Second Baku', the first discoveries in this area took place as early as 1932.[56] Expansion of the region was delayed, however, by a shortage of the proper drilling equipment. It was only after the Second World War that petroleum in the area was produced in large quantities and that the region, particularly at the giant field of Romashkino, came to outproduce Baku.[57]

CONCLUSION

What is unique about matters affecting the Soviet petroleum industry prior to the Second World War was that so little was unique. Much of what happened had already occurred in the pre-revolutionary years or, as we shall see, will be repeated after the Second World War. For that reason it is worth summarizing what happened so that we can more easily note the similarities when they recur in the pages ahead.

Foreign help was very important to the Russian petroleum industry prior to the revolution, as well as before the Second World War. That includes technological assistance at the drilling, extracting, and refining stages.Nor did the Soviets refrain from seeking foreign help to facilitate the foreign marketing of their petroleum. Often that meant selling to companies like Standard Oil or Shell, so that they could do the distributing. In other instances, it meant joining together with a local company to form a joint Russian-local venture not only to handle wholesale distribution but on occasion to operate retail filling stations as well. The concept of trading with multinational enterprises or even creating their own multinational network evidently posed no ideological hurdle for the Soviets. Nor, for that matter, was politics much of a barrier. The Soviets managed to sell their petroleum to Mussolini's fascists and Hitler's nazis, even when decency and the popular political line should have precluded such action. The politics of ideology was seldom allowed to stand in the way of the principle of profit.

One justification for foreign help was the periodic fear that Russian reserves might run out and that output was being used ineffectively. The same fears were expressed prior to the revolution, and as we shall see will recur. The concern was an important one because increased production was essential, not only because of the need to supply domestic demand, but because of the role petroleum played as an earner of foreign currency. At its peak, in 1932, petroleum accounted for 18 per cent of foreign earnings. That depression year also saw the Soviet Union export more petroleum than did the United States, and probably more than anyone else in the world. To make possible exports of this magnitude, the Soviet Union set aside 29 per cent of its production for this purpose, apparently without seriously curtailing its domestic consumption. Again, this was a pattern that was set before the revolution and one that, like so much else, has continued into the 1970s.

NOTES: CHAPTER 2

1 Robert W. Tolf, *The Russian Rockefellers* (Stanford: Hoover Institution Press, 1976), p. 40.
2 G. Segal, 'The oil and petrochemical industry in the Soviet Union' (London: mimeo., undated), p. 1.
3 Tolf, pp. 41–2.
4 Iain F. Elliot, *The Soviet Energy Balance* (New York: Praeger, 1974), p. 70.
5 ibid.
6 Segal, p. 2; V. I. Pokrovskii, *Sbornik Svedenii Po Istorii i Statistike Vneshnei Torgovli Rossii* (St Petersburg: Department Tamozhennykh Sborov, 1902), Vol. I, p. 217.
7 Pokrovskii p. 217.
8 Elliot, p. 70; Tolf, p. 44.
9 Tolf, p. 40. This is an excellent study of Nobel's life and work. Much of the description of Nobel's activities in the remainder of this chapter is taken from this book and the study by Pokrovskii.
10 Pokrovskii, p. 218.
11 Tolf, p. 55.
12 Pokrovskii, p. 219.
13 Segal, p. 3.
14 Pokrovskii, p. 220.
15 Tolf, p. 97.
16 ibid., p. 95.
17 Pokrovskii, p. 221.
18 Tolf, pp. 48, 71–2, 99.
19 Pokrovskii, p. 221.
20 US Bureau of the Census, *Historical Statistics of the United States, Colonial Times to 1970*, bicentennial edn, Pt I (Washington, DC, 1975), p. 597.
21 Pokrovskii, p. 215.
22 Tolf, p. 43.
23 ibid., p. 142.
24 Pokrovskii, p. 215; Peter Lyashchenko, *History of the National Economy of Russia to the 1917 Revolution* (New York: Macmillan, 1949), p. 689.

25 Zbynek Zeman and Jan Zoubek, *Comecon Oil and Gas* (London: *Financial Times*, 1977), p. 9; Tolf, pp. 188–92.
26 ibid., pp. 192–3.
27 Pokrovskii, p. 217; Segal, p. 2.
28 Lyashchenko, p. 632.
29 Exports in 1913 however did slightly exceed those in 1905 by 3,000 tons.
30 Trudy statisticheskogo otdelenie, Departament tamozhenny sbor', *Obzor' Vneshnei torgovli Rossii za 1901* (St Petersburg: Departament tamozhenny sbor', 1903), p. 9 (hereafter *Vneshnei torgovli Rossii* and the appropriate year); *Vneshnei torgovli Rossii 1915*, p. 8.
31 A. V. Venediktov, *Organizatsiia gosudarstvennoi promyshlennosti v SSSR* (Leningrad: Izdatel'stvo Leningradskogo Universiteta, 1957), p. 185.
32 Tolf, p. 215.
33 ibid., p. 217.
34 Segal, p. 5.
35 Anthony C. Sutton, *Western Technology and Soviet Economic Development, Vol. I, 1917 to 1930* (Stanford: Hoover Institution Press, 1968), p. 41; Tolf, p. 222.
36 Sutton, p. 18.
37 ibid., pp. 19, 30, 122; the Japanese did not come in until 1925.
38 ibid., p. 37.
39 Anthony C. Sutton, *Western Technology and Soviet Economic Development, Vol. II, 1930 to 1945* (Stanford: Hoover Institution Press, 1971), p. 17.
40 ibid., pp. 10, 23.
41 ibid., p. 82.
42 Venediktov, p. 185.
43 Segal, p. 16.
44 *Foreign Trade*, September 1967, p. 18.
45 Sutton, Vol. I, p. 41.
46 *Foreign Trade*, September 1967, p. 17.
47 Ministerstvo vneshnei torgovli SSSR (hereafter MVT SSSR), *Vneshniaia torgovlia SSSR za 1918–1940* (Moscow: Vneshtorgizdat, 1960), p. 657 (hereafter *VT SSSR* with the appropriate year).
48 ibid., p. 528.
49 Sutton, Vol. I, pp. 16, 40.
50 ibid., p. 40.
51 Tolf, p. 224; Segal, p. 8.
52 *VT SSSR*, 1914–40, p. 94.
53 *Journal of the USSR Trade and Economic Council*, July/August 1976, p. 7.
54 Robert W. Campbell, *The Economics of Soviet Oil and Gas* (Baltimore: Johns Hopkins University Press, 1968), p. 27.
55 Segal, p. 12.
56 Elliot, p. 72.
57 ibid., pp. 74–5, 89; Campbell, 1968, pp. 124–5.

3

*The Post-1939 Period to 1973 –
Internal Developments*

To Hitler, Russia stood for wheat and petroleum. But Hitler's information was dated. As we saw, once under Soviet control Russia's grain surpluses diminished rapidly so that, whereas pre-revolutionary Russia had managed to export 9 million tons of wheat in 1913, the most the Soviets could muster prior to the Second World War was 5 million tons in 1931, and to do that they had to starve their own people.[1] The most grain the Soviets have ever been able to export was 8·6 million tons in 1971. They could only do this, however, by simultaneously importing 3·5 million tons of grain.[2]

But if the bread basket of Europe was not as full as it once was, the oil wells were as attractive as ever, and Hitler sought to capture the Baku fields. To his dismay, German troops did not quite reach Baku, but they did capture the Grozny fields in the north Caucasus. Even here, however, the fields were so badly destroyed that Hitler was unable to derive much benefit from them, but he did deny their use to the Soviets. Moreover, supply routes from Baku to the north were adversely affected, so that the Soviets had a hard time maintaining their fuel supply. They were helped to some extent by Lend-Lease oil shipments of 2·7 million tons of petroleum from the United States. None the less, by the time the war ended, many Soviet oil fields had been destroyed, so that in 1946 Soviet oil production had fallen to 22 million tons from the 1940 peak of 31 million tons.[3]

THE VOLGA-URAL REGION

To help in the reconstruction of both the fields and refineries foreign help was called in, including $1 million worth of equipment confiscated from Rumania.[4] Most of the Soviet effort was directed at reconstituting and expanding the traditional Baku area, but gradually they began to move towards the meagerly developed Volga-Ural region. Although exploration in the newer area predates the revolution, no oil was found until 1929.[5] Even then not much happened, so that by the time the

Second World War started annual output was not quite 2 million tons a year. Some major finds in Devonian deposits during the war in 1944 were made, but serious drilling work only began in 1955.[6] Until then, the main emphasis on drilling continued to be restricted to the Baku region.

Since the war had lasting effects on petroleum production in the Baku region, the Soviets were fortunate that the Volga-Ural region was available to them. Despite considerable drilling effort, output in Azerbaidzhan never fully recovered. Even in 1966, the postwar peak, Azerbaidzhani oil output was unable to equal the 22·2 million tons pumped in 1940. But as new fields were discovered in the Volga-Ural region, output rose so rapidly that the area soon out-produced Azerbaidzhan, and by 1949, total output in the USSR surpassed the previous level of production. It has risen ever since. (See Table 2.3.)

Output in the Volga-Ural region continued to increase until about 1970. The field at Romashkino in the Tatar ASSR was for some time thought to be the largest in the world. But after 1965 output in this region began to fall sharply.[7] The response was to seek some way to enhance secondary recovery. The initial solution was to inject water into the wells in order to restore the pressure needed to facilitate the extraction of petroleum. In some cases water injection was only partially successful. Occasionally it made matters worse. Even where it worked, after a time so much water was injected that it became even more difficult to extract the petroleum the water was intended to flush out. Special pumps were required, and often more water was taken out than oil. Since this is the main technique the Soviets have used to enhance petroleum output, they must somehow find another way to increase production. Until they do, they have a serious problem on their hands. This is also recognized by the CIA. The fall-off in output because of water injection is one of the main underpinnings of the April 1977 CIA report.[8] In the CIA's view, unless the water injection problem is solved yields per well all over the country will sooner or later drop off sharply as they did in the Volga-Ural region twelve years earlier.

WEST SIBERIA

The Soviets were spared the need to solve the problem in the Volga-Ural region because just as output per well there began to fall the slack was taken up by the opening of brand new areas in west Siberia. Moreover, whereas it took from 1929 when the first oil was struck in the Volga-Ural region until the late 1940s for production to reach a meaningful level, the lag between discovery and production in west Siberia was much shorter. Although exploration for liquid energy in the region began before the Second World War, the first find occurred by accident in west Siberia in September 1953.[9] A drilling team was delayed while sailing up the Ob river near the town of Berezov. On the spur of the moment it made a test boring and found gas in what it came to call the Berezovskoe gas field. It was seven more years, in 1960, however, before the first oil was discovered in a Jurassic zone near Shaim on the River Konda, a tributary of the Ob and Irtysh. The 'super-giant' field in a Cretacean level at Samotlor, about 500 miles to the east, was discovered in 1965, and the first commercial-scale well completed in April 1968.[10] Whereas it took almost twenty years for the Volga-Ural fields to move from discovery to production, it took only eight years in the west Siberian Tiumen region. By 1970 production had already reached 31 million tons; by 1975 it amounted to 145 million tons; and in 1977, about 210 million tons.[11]

Notice the pattern here. The output in west Siberia seemed to compensate for the drop in productivity in the Volga-Ural fields in the late 1960s and early 1970s, just as the coming on line of output in the Volga-Ural regions in the late 1940s and early 1950s offset the declining output of Baku. So far, each time output in one major field has slackened, the Soviets have come up with a new find. It would be nice if they could continue in this leapfrog manner. But this type of solution to their production problem also has a negative side. It has served to postpone the time when the Soviets must face up to the need to use their resources more effectively.

THE SOVIET PLANNING, PRODUCTION, AND INNOVATION SYSTEM

To comprehend why it is so difficult for the Soviets to solve their efficiency and productivity problems requires an understanding of their planning and incentive system and the special peculiarities that affect the Soviet raw materials and petroleum industries. In fairness it should be pointed out that backwardness in innovation is not an affliction brought on solely by the Russian Revolution. The revolution seems to have compounded the problem, but the malaise was there before 1917. As we saw when discussing drilling technology in Baku in the nineteenth and early twentieth centuries, Russia's existing technique continually lagged behind developments in the West. Invariably foreigners had to be called in either to import more advanced technology or to run actual concessions.[12] Such lags are not necessarily characteristic of all pre-revolutionary and pre-Five-Year Plan technology, but they are widespread enough to cause suspicion that something deeper than a poor incentive system is the explanation. It may be that Russia was too remote to be affected by the West European Renaissance, or that Napoleon never stayed in Russia long enough to bring with him the reforms of the French Revolution and their emphasis on scientific enlightenment and rationality, or that widespread literacy was lacking until midway into the twentieth century, or that the oppressive legacy of authoritarian and rigid governments before and into the revolution snuffed out all initiative. It could be a mixture of all these factors. Whatever the exact explanation, there is no doubt that Russian culture and history seem to have mixed with the Soviet system of central planning and economic incentives to stifle creative thinking, at least in the economic and technology spheres.

An example that illustrates how deficiencies in the Soviet process of producing, planning, and innovating have affected the development of the Soviet petroleum industry is the Soviet development of the turbo-drill. In a perceptive analysis, Robert Campbell of the University of Indiana explains why the innovation-shy Soviet petroleum engineers were prodded into developing this unique process.[13] To have pursued the conventional path being followed by their Western counterparts, the Soviets would have had to have high-quality steel-pipe subcomponents. But while it is true that innovation is a rare phenomenon in the USSR, it was simpler to invent some new technique than to up-end the whole Soviet production process in order to achieve the necessary pipe quality. Without a revolution or major reform in Soviet planning and incentives, there seemed to be no way to gain access to the high-quality products needed for drilling that are taken for granted in other industrialized societies. Since such a thorough shake-up of the economic system seemed not only impractical but impossible, the only other way to solve the critically important drilling and production task was to find some new approach. This led to the turbo-drill.

Poor-quality pipe is symptomatic and at the same time the cause of the problem. To drill effectively using conventional techniques, the driller must have good-quality pipe. The pipe must be able to withstand the increasing tension and pressure as the drilling goes deeper. With the poor-quality steel pipe, breakdowns, cracked pipe, and tool-joint failures are endemic.[14] This means not only an increased need for replacement pipe but lost time spent on repairs and lifting and lowering the portions of the pipe string that remain intact. With the type of pipe available to the Soviets at the time, it was only possible to drill down to 2,000 meters.[15] In 1950 that was the depth of almost 90 per cent of Soviet wells. Though inefficient, it was adequate for the drillers in the Baku region. The Soviets were able to satisfy their needs, albeit with a good deal of waste.

While shallow wells may have been suitable for Baku, they were of no use in the other regions where oil and gas were located at deeper depths. Moreover, the use of the rotary drill process with low-tensile pipe meant that the drillers could only reach the 2,000 meter depth when the ground was soft and the rock not too hard. But, as Campbell points out, Soviet drill pipe at best is made from what in the United States is considered grade D and some grade E steel.[16] In the United States, however, drillers restrict themselves to the use of grade E steel and pipe or even higher grades. Why does the Soviet Union, as the world's largest producer of steel, not produce higher quality steel?

As long as the Soviet system places stress on quantity rather than on quality of production, the Soviet manager has little or no incentive to produce the highest-grade steel. His pay depends not on producing high quality but on producing as much quantity, usually measured in weight, as possible.[17] For the most part, the Soviet manager is not concerned whether his product is sought after in the market place. It is usually enough that his product is produced and transferred from the factory floor. Until recently at least, once that happened, everyone was entitled to share in the enterprise bonuses.

To ensure that such bonuses would be forthcoming, the factory manager devoted virtually his whole effort to ways of increasing production. Over time, Soviet managers have developed a fine-honed sense of just how this should be done. Those who did not succeed were discarded along the way and did not reproduce. But Soviet economists soon discovered that a single-minded devotion to quantity led managers not only to ignore quality and variety but to dispense with them. For example, every time a machine was shut down to change size or to improve the process, there was less time available for production. Occasionally a change in process might lead to faster or improved production, but there was always the uncertainty that the innovation would not succeed and production would not increase. At the same time, however, there was always the certainty that production would be curtailed at least temporarily. Since the odds were against taking such risks, quality improvement inevitably suffered.

For the oil drillers, this meant that the steel manufacturers they depended upon had no incentive to produce or even contemplate producing the higher-grade qualities of steel.[18] In his colorful way, Khrushchev put it vividly when he complained: 'The production of steel is like a well traveled road with deep ruts; here even blind horses will not turn off, because the wheels will break. Similarly, some officials have put on steel blinkers; they do everything as they were taught in their day. A material appears which is superior to steel that is cheaper, but they keep on shouting "steel, steel, steel!".'[19] While this attack was delivered in the context of the planners' inability to switch to new, more sophisticated and innovative industries like electronics, computers, and chemicals, Khrushchev's complaint was equally valid when addressed to the need for qualitative improvements within the steel industry itself. Innovation was disruptive and therefore to be discouraged.[20]

Periodically efforts have been made to break out of this conundrum and put more pressure on the manager to improve quality. The most recent and notable effort occurred in the 1960s with what was described in the West as the 'Liberman reforms,' named after Evsey Liberman, one of the reforms' chief proponents. Under this experiment, premiums to factory managers were to be made dependent on sales to the customer, not on production for the warehouse. Sales and profits were to measure managerial and productive ability, not simply the fulfillment of production quotas. To the extent that plant managers found that they had to concern themselves with customer needs and tastes, the reformers hoped that quality would improve. For this to happen, however, there would have to be a relaxation of the tautness that had been all but a permanent feature of the Soviet economic system. No matter what reforms might be decreed on paper, if Soviet buyers found themselves, as before, with a shortage of suppliers and the continued existence of the traditional seller's market type atmosphere, they would find that they had only very limited leverage on producers. Under the circumstances, nothing would be done to change the 'take it or leave it' attitude of the producers.

The success of the reform was also contingent on some basic changes in the pricing system. If markets and demand and supply were to serve as barometers of need, the manufacturer should have the power to lower or increase prices. Such power would also be a way of providing rewards for innovation and risk-taking. As long as prices are rigidly controlled from the center, the manager has to spend as much time negotiating with the bureaucracy for an appropriate award for his innovation as he spends on innovation itself. Inevitably this detracts from the process of improving quality and innovation, since having to seek permission from the bureaucracy entails not only large amounts of time and effort but also the strong possibility that such permission will be denied.

Most important, to be implemented the reforms required the acquiescence and support not only of the managers but of senior ministry authorities. Ultimately, such support was not forthcoming. The reasons are not hard to understand. There were neither managers nor ministers equipped to deal with this brave new incentive world. The Darwinian selection process had bred them out of the system. Those who survived or throve were those who knew how to make the most out of a system that stressed the production of ever more of the same thing. The thought of having to be responsive to customer tastes and changes in quality and process was an upsetting prospect. Moreover, the reforms were introduced in a piecemeal fashion that made them difficult to implement. Only a few factories at a time were given the freedom to alter output and prices, so as to maximize profit, not output. Inevitably they found that their new freedom was negated when they could not locate suppliers who were willing to match their willingness to innovate. What good is it to attempt to produce more modish suits, for example, when the fabric manufacturers are producing the same old type of cloth and have no desire or incentive to produce any of the new synthetics? Or suppose you do find someone willing to produce such cloth; what do you do when you find that the old sewing machines are not equipped to sew this new material, and that the sewing machine manufacturers are not willing to upgrade their machines because they are on the old planning system that throws up disincentives in front of any innovation?

To illustrate the dilemma, Soviet economists used to describe how Soviet traffic authorities once went about improving Moscow's traffic problems. In an effort to speed up the flow of traffic, the city planners sent a specialist to London to study how the British were handling their traffic problem. Impressed by what he saw, he returned to Moscow with a simple plan. 'Everyone in London,' he said, 'drives on the left side of the road. We must do the same.' After much discussion, the planners decided that they would make the needed reform and that as of 1 January all traffic would be shifted from the right to the left side of the street. But as the day approached, some of the planners began to have second thoughts. Such a change introduced all at once might be too much for Moscow drivers to assimilate, particularly since drivers in some of the vehicles would have poor visibility if they had to drive in a left lane from the left side of the vehicle. Again there was much discussion, and ultimately a compromise was reached. As planned, on 1 January all trucks, buses, and taxis would be switched to the left side of the road; but the less experienced drivers of automobiles would continue on the right side as before until 1 July!

When the 'Liberman Reforms' were promulgated in a similar step-by-step manner, whatever slim chances of success there might have been were dashed. Even those few managers and ministers who supported such experimentation found it to be counter-productive. The managers found they could not really bring about the necessary change because they lacked the power they needed to do what they saw they should do. Moreover, they depended on other managers who had no power at all. As for the ministers, they found that they were still being held responsible for increasing the quantity of goods being produced even though they had been instructed to sit back and turn over some of the powers to the managers. Finally, in 1968, just as the reforms were reaching a critical stage, spring and a somewhat similar set of reforms came to Czechoslovakia, setting off what seemed to be unrestrained experiments in democracy and economic tinkering. Horrified by what they came to believe would be the inevitable consequence of like reforms in the Soviet Union, Soviet leaders decided to end the experiment not only in Czechoslovakia, but also in the Soviet Union. One result is that Soviet oil men have been waiting for a long time for high-quality steel and pipe to be produced in the USSR.

THE TURBO-DRILL

Having concluded, therefore, that the changes in the economic planning and incentive system (needed to produce high-quality pipe) would be slow in coming, if they came at all, Soviet oil planners in the 1940s, 1950s, and 1960s realized they would simply have to do without domestically produced high-grade Soviet pipe. Of necessity they would have to find some other way to locate their oil and lift it out of the ground, preferably without having constantly to rotate their poor-quality drilling pipe.

The Soviets were not the only ones to seek a substitute for the rotary drilling method. Several American engineers had also conceived of the same approach.[21] One alternative was a hydraulic or turbo system that would use a steady flow of water to move the drill bit on the bottom of the string of pipe. This way the pipe would remain fixed, and therefore extra-strength pipe was not so important. The only movable elements in the system would be the water and drilling mud, which moved a turbine at the bottom of the pipe, which in turn activated the drill bit. Yet while the theory of the process made sense, no one could work out the actual operations. Soviet engineers began their first efforts with such a turbo-drill in the 1920s. Some progress was made in 1934, but not enough to solve the basic engineering challenges that were involved. But since the only other alternative seemed to be continuous reliance on imported Western pipe, development work continued. During the war, the experiments were moved from Baku to the greater safety of the Volga-Ural region.[22] In retrospect this proved to be a stroke of luck, since the turbo-drill, which was about to become a reality, was not suited for Baku, but was ideally suited for the Volga-Ural region. The hard rock in the Volga-Ural region lent itself nicely to penetration by the turbo-drill, at least up to 2,000 meters in depth. This was fortunate since the Volga-Ural rock all but blocked penetration when the rotary drill process was used with Soviet pipe.

However, once the turbo-drill had proven itself, its use spread rapidly. By 1956 it accounted for 86 per cent of all Soviet drilling.[23] Since no one else, including the Americans, was able to master the engineering involved, the Soviets even began to export as much as 6 per cent of their output.

But while the turbo-drill may have been well suited for the Volga-Ural region, it was of little use at greater depths or in soft rock.[24] As explained by Campbell, efficient use of the turbo-drill system requires high-speed operation of the turbine and drill bit. Unfortunately, however, the higher the speed, the faster the bit wears out, especially Soviet bits, which deteriorate much faster than American-made bits. According to one American manufacturer, while his bits penetrate 6 to 8½ meters a day, Soviet-made bits carve out only 1 meter a day.[25] Even more serious, the average life of the Soviet drill bit gives out after about 21 meters.[26] American manufacturers claim that American drill bits last five to twenty times longer than their Soviet counterparts. Still, the problem is not one of having enough drill bits. After all, the Soviets produce more drill bits than we do. The Soviets produce 1 million drill bits a year, compared to production of 400,000 for all the rest of the world.[27] But that is part of the problem. As with the steel pipe, the Soviet drill bit factory is designed to put out quantity, not quality. Thus, a typical Soviet factory produces 255,000 bits a year, but only two models.[28] In the United States, a typical factory produces only 70,000. In part this is because production is frequently interrupted to allow the firm to tool up for the 600 models it offers.

The inappropriateness of the poor quality of Soviet drill bits explains why Soviet drillers take so long, and why their drill bits wear out so fast. Moreover, every time the drill bit wears out, the whole string of pipe must be taken out of the ground. Obviously, the deeper the hole, the longer the process takes. So not only does the Soviet drill bit in the turbo-drill process carve out less per day, but it wears out faster. As a result, drilling in the Soviet Union is much slower than it is elsewhere in the world. According to one estimate, in a thirty- to thirty-five-hour period of time, only four to five hours, or about 15 per cent of the time, is devoted to drilling; the remainder of the time is spent on withdrawing or reinserting the pipe string and replacing the drill bit.[29] On the average, it takes the Soviet driller five times longer to drill a meter than it does his American counterpart.[30]

In 1965, as yields began to fall in the Volga-Ural region, the Ministry of the Petroleum Industry officials were confronted with a serious challenge. As we saw in the early and mid-1960s, petroleum had been discovered in the west Siberian region around Tiumen. But the turbo-drill was ineffective at the greater depth necessary to find deposits in these new regions.[31] Moreover, these new fields were located at considerable distance from existing processing centers and pipelines. New measures were required to find and drill these new fields and build the pipelines to carry the petroleum and gas to processing and consuming centers.

One approach to the challenge was to improve the operation of the turbo-drill. Some claims were made that the turbo-drill digging capability had been extended to 3,500 meters.[32] However, as long as the drill bit quality was so poor, it was difficult to reach such depths with domestically produced equipment. According to CIA estimates, it takes more than a year for Soviet oil men to drill 3,000 meters. By world standards, this is too long a period of time. Certainly, a technological breakthrough is essential if the Soviets are to make efficient use of their large stock of turbo-drills.

Another approach has been to find some alternative to the turbo-drill. Whereas turbo-drills in 1956 accounted for 86 per cent of all drilling equipment, by 1975, the ratio had fallen to 80 per cent.[33] For example, one of those alternatives involved the use of an electro-drill where the bit is driven by an electric motor.[34] Because it was thought that it could drill to depths of 5,000 meters, it was initially regarded with great expectations.[35] But Soviet manufacturers experienced considerable difficulty in producing it. Thus it did not become a viable option.[36]

In lieu of the turbo-drill and electro-drill, the Soviets have been forced to fall back on the rotary drill. After concluding that they could not solve the innovative barrier, and produce the necessary steel pipe themselves, they have finally decided to import turnkey factories from the United States, Germany, and Japan that will produce higher-grade steel from the day they are open. In other words, they have given up the effort to create factories that would be designed and introduced by the Soviet Union's own steel enterprises. The Soviets are currently building two or three such foreign high-grade steel factories with the hope of solving the steel quality deficiency that affects not only pipe manufacturing, but several other fields as well.

In the interim, however, the Soviets have had to import pipe and pipe casing. Obviously not all of the imported pipe is used for the rotary drilling rigs. By far the bulk of imported pipe goes towards construction of petroleum and gas pipelines, but since that is also relevant to our study, it is worth mentioning here. In fairness it should be remembered that the United States also has had to import large quantities of pipe recently from Japan for the Alaska pipeline, since we, like the Soviets, lack the proper pipe-making equipment. But, unlike the Soviets, we at least have the steel that is needed to make the pipe.

Soviet pipe imports are worth fussing about, because their magnitude in some recent years, for both drilling and pipeline, has been large. Soviet pipe imports constituted about 15 per cent of all Soviet hard-currency imports. In addition, significant quantities of pipe were imported from some of the East European countries, like Rumania and Poland.

Much of this pipe, from both the hard-currency countries and Poland – but not from Rumania – did not cost the Soviet Union any out-of-pocket money. It was delivered in return for promises of future deliveries of Soviet oil and gas. In a sense it was a form of barter. Yet how much easier and cheaper it would have been for them if their industrial system had been more innovative and responsive, and the Soviets could have produced the pipe themselves.

RAW MATERIALS EXTRACTION AND THE SOVIET SYSTEM

Troubling as pipe production inadequacies have been, and continue to be, they are not the only shortcoming in the Soviet economic system that adversely affects the petroleum and gas industries. The whole history of raw materials extraction is plagued by a series of problems that have caused enormous waste over the years and have done little to reverse the pre-revolutionary pattern. It may not be too much of an exaggeration to suggest that instead of increasing the output and efficiency of raw material extraction, the Soviet planning and economic system has occasionally, if not generally, been the source of needless waste and confusion. Making such an assertion does not mean that mineral extraction and usage in the non-communist world is waste-free or even efficient. This we saw during the energy crisis of 1973 and 1979. Nevertheless, the Soviet system by comparison is not much better and in some instances may be worse.

Since many of the problems affecting Soviet planning and incentives are systemic, we should not be surprised to see the same problems arise in the extraction and production of other raw materials. However, in addition to the inefficiency that stems from the planning system, adherence to Marxist ideology often creates additional waste. Normally, in the Soviet Union, ideology is kept in its place and not allowed to complicate or hamper economic production. Soviet economists have become adept at rationalizing awkward ideological injunctions. However, for some reason this is less true in raw material exploitation. For instance, on 2 September 1930 the Soviets passed a law abolishing payments and rent on land to be mined.[37] As a result, until recently the Soviets treated minerals in the ground as free goods. By contrast, Soviet managers had to pay for the labor they used as well as the capital goods. The result was readily predictable. In the words of Academician Khachaturov, the enterprise 'prefers to make more economical use of its capital even if it means neglecting natural resources.'[38]

As opposed to the American or capitalist mine operator who has to pay for the raw materials he mines in the form of either rent, royalties, or the purchase of land, the Soviet mine operator is provided with raw materials free of charge. Thus he bears none or almost none of the land costs that a capitalist miner must face. As a result, the Soviet mine operator will not attempt to exploit the mineral deposits as extensively as will his capitalist counterpart. The Soviet manager is more likely to dig or drill and then run. The capitalist miner or driller is more likely to stay and take out a much larger percentage of the deposit.

Like miners all over the world, the Soviet miner takes out the richest ore first. As the richest ore diminishes, the miner's costs per ton of output begin to rise. The Soviet miner focuses only on his marginal and variable costs. As these costs begin to rise, the Soviet miner begins to look for another, easier and cheaper mine to exploit. Like miners in the capitalist world, the Soviet miner does not have to worry about the costs already put into the original mine. Both in the Soviet Union and the United States, all past capital costs are no longer a concern. 'Bygones are bygones.' Economists and miners do not cry over already 'spent capital' or spent land acquisition. However, here the similarity ends. Unlike his American counterpart, the Soviet miner is also spared the need to worry about his future land acquisition or raw material acquisition costs.[39] In effect, because of the peculiarities of the Soviet pricing system, the new mining site is a free good to the Soviet mine operator. In addition, until 1 July 1967, the geological exploration cost was likewise something the Soviet mine operator did not have to cover. Then 25 per cent of the geological costs were included as a cost of production. Even now, not all of these costs are levied. As of mid-1976 miners were charged only 60 per cent of all their geological exploration costs.[40] Therefore, when the Soviet mine operator finds that his marginal costs of operation at the old site exceed the average costs of labor and capital plus the average cost of moving to the new site, he will move.

In contrast, the capitalist miner has more costs to contend with and, therefore, he is more likely to stay in place longer and he needs to extract more in order to make a profit. The difference is due to the fact that, in addition to all the costs the Soviet miner has to worry about, the American miner also has to include a fixed cost in the form of a new purchase of land. Thus, unlike the Soviet miner, the capitalist miner has to make allowance for new average land costs per unit of output as well as the full geological costs before he contemplates moving. Thus, the costs of operating at the new site will appear to be higher to the American mine operator than they are to the Soviet mine operator. In contrast to the American miner who sweeps his mine clean, the Soviet miner is more likely to ignore the edges and the harder-to-reach corners of his deposit. Also he is more likely to leave pillars of coal and other raw materials standing to hold up the ceiling rather than bring in other, seemingly more expensive, forms of support. The natural pillars left in the USSR often contain more than 20 per cent of the mine's ore or coal.[41] Potassium salt pillars often amount to as much as 50 per cent of the potential output.[42] Recognizing these tendencies, Soviet economists like Federenko and Khachaturov have been arguing for the introduction of a rent and raw material charge. As Khachaturov sees it, 'if the enterprise has to pay for natural resources, it will treat them as carefully and economically as productive capital.'[43]

To the extent that the pricing mechanism fails to reflect the full economic costs at an early stage of production, it is almost certain that such distortions will be carried throughout the rest of the economic system. Thus, raw materials tend to be underpriced in the Soviet Union. This in turn induces consumers of raw materials to use more than they otherwise would. Accordingly, the Soviet Union expends more fuel per kilowatt of electric power and per ton of open hearth steel smelted, and more metal per unit of engine power than the United States does. In the Soviet machine tool industry, for example, over 25 per cent of all the rolled steel used is discarded as scrap.[44] Given the planning system with its emphasis on output and the tendency to understate or ignore the true costs involved, it was inevitable that there would be waste and inefficiency in Soviet mining practices.

This waste is translated into Soviet extraction ratios that are very much lower than those that prevail in the non-communist world. Admittedly, the whole concept of a norm for extraction rates leaves something to be desired. What can be extracted at any given time in any given country keeps changing in response to changes in prices and technology.[45] Seldom will two engineers, economists, or geologists agree exactly as to how much of a mine or well can be extracted. Yet, whatever the precise concept, Soviet economists, engineers, and geologists have an approximate idea of how they are doing in relation to what is technologically and economically feasible. They complain repeatedly that Soviet mining and drilling practices are needlessly wasteful. In contrast to the American experience, where recovery rates in coal mines, particularly strip mines, are often 90 to 100 per cent, in the Soviet Union the figure is frequently only 70 per cent.[46] The recovery rate of mica is as low as 10 per cent, while the recovery of potassium salts reportedly is 40 to 50 per cent of what could be extracted. The extraction rate of ferrous and non-ferrous metals in the Soviet Union is not much better. At the Krivoi Rog mines it is only 54 per cent.[47] This general tendency is confirmed by the Soviet economist and member of the Academy of Sciences T. Khachaturov, who reports that often 40 to 50 per cent of the solid minerals that could be mined is left in the ground.[48]

Even more pertinent for our purposes, the petroleum recovery ratio for west Siberia is reported to be only 30 to 40 per cent.[49] One specialist writes that the figure may sometimes be as low as 25 per cent.[50] What makes this particularly distressing is that this is below the Soviet national recovery goal of 41 to 45 per cent.[51] However, the 41 to 45 per cent figure is unrealistically high. The figure in the United States is about 32 to 33 per cent, and since the Soviets presently are woefully behind in the technology of enhanced recovery, they are unlikely to attain such relatively high recovery rates.[52] But for our purposes it is enough to notice that, however the recovery index is measured, the Soviets themselves are fairly unhappy about the fact that yields in west Siberia are one-third to one-eighth below the national goal. According to the economist N. G. Feitelman, their goal is a recovery index of 80 to 90 per cent, which they insist would make the Soviet recovery of petroleum the most efficient in the world.[53] However, in the eyes of some Western oil men, such a high rate of recovery is a virtual impossibility. The point is, however, that the Soviets think it conceivable that recovery yields in west Siberia could be as much as three times more than they are at present and more than double what they are in the country as a whole.

The wasteful flaring of the natural gas produced as a by-product of petroleum has also been of special concern to Soviet economists. For each ton of oil extracted there is usually an associated 150 to 200 cubic meters of gas.[54] For a time as much as 40 per cent of this gas was flared.[55] As recently as 1975, 20 billion cubic meters or 7 per cent of the country's total output was wasted this way.[56] It should be pointed out that for many years gas was similarly flared in Iran and Saudi Arabia. That was tolerated as long as both countries did not concern themselves too much with industrialization. But as first Iran and then Saudi Arabia began to industrialize, both sought to halt the waste involved in burning up a valuable raw material and took steps either to sell the natural gas or to utilize it domestically. The Soviets, however, from the beginning have had industrialization as a prime goal. Therefore their flaring of natural gas was extremely wasteful.

The planning system also is ill-suited for locating new deposits. Since planning targets are usually spelled out in terms of some physical measure, for those in agencies like the Ministry of Geology whose work involves drilling, the most reasonable index seemed to be the number of meters drilled. The more meters drilled, the better the performance – or at least, so one would think. Unfortunately, Soviet geologists soon discovered that the deeper they dug the longer it took them and the less likely it was that they would fulfill their plan. As a result, the geologists quickly developed the practice of drilling shallow holes. As an article in *Pravda* put it, 'Deep drilling means reducing the speed of the work and reducing the group's bonuses.'[57] A description of the area sounded more like a smallpox than a mining report. 'In some places, the land is becoming increasingly pitted with shallow, exploratory holes drilled in incessant pursuit of a larger number of total meters drilled.' It comes as no surprise, therefore, that 'there are geological expeditions in the Republic of Kazakhstan that have not discovered a valuable deposit for many years, but are counted among the successful expeditions, because they fulfill their assignment in terms of meters. The groups that conscientiously "turn up" deposits are often financial losers.'

Moreover, even if a deposit should be found, the drillers from the Ministry of Geology bear no responsibility for determining the size of the deposit. Consequently, the actual producing ministries must also maintain their own drilling units. In some instances there may be as many as three separate drilling agencies duplicating one another's work.[58] Undoubtedly it would be much more efficient if it were possible to base the drilling team's compensation on the amount of raw materials actually recovered, but this has been resisted by the planning agencies and the Ministry of Geology which fears it would disrupt its planning procedures. As so often happens, they have confused the means, that is how many meters are drilled, with the end, how much oil is found.

There is another way in which the Soviet planning system creates institutional blockages to a more efficient utilization of mineral deposits. Responsibility for mining in the Soviet Union is usually divided up among several large ministries or state committees. For instance, there is the Ministry of Ferrous Metallurgy, the Ministry of Non-Ferrous Metallurgy, the Ministry of the Coal Industry, the Ministry of the Chemical Industry, the Ministry of the Gas Industry, the Ministry of the Petroleum Industry, and the Ministry of the Construction Materials Industry. Unfortunately, nature does not always break itself up into the same neat and precisely defined categories. Thus a valuable mineral such as apatite is often mixed in with iron ore in the ground. Yet because the Ministry of Ferrous Metallurgy is only accountable for iron ore output, not apatite, a million tons a year of this mineral ore was junked at the Kovdor mine by the agencies of the Ministry of Ferrous Metallurgy rather than used for fertilizer where it has a high value.[59] Similarly, the gravel extracted with iron ore at Krivoi Rog and Kursk is also tossed aside as waste instead of being used for construction. In the same way, the Ministry of Ferrous Metallurgy also rejects as slag significant quantities of valuable minerals such as vanadium, titanium, nickel, chrome, cobalt, and copper, and even platinum.[60] In its turn, the Ministry of Non-Ferrous Metallurgy discards valuable quantities of ferrous metals.[61] Some of this waste could probably be eliminated by more co-ordinated planning, but some of it will not be eliminated until there is a drastic revision of economic incentives.

The failure of the Soviet bureaucracy to divide itself up in the same way as nature also helps to explain why so much natural gas is flared. Virtually none of the flaring is done by the Ministry of the Gas Industry. Most of it is done by the Ministry of the Petroleum Industry, which produces the gas as a by-product with petroleum as its main concern.[62] Therefore, until recently at least, the plan fulfillment efforts of the Ministry of the Petroleum Industry were little affected by what happened to the by-product, natural gas, with which it ended up.

PRICING

Our list of structural shortcomings must also include mention of the pricing system. Not only do prices cause distortion, because inadequate allowance is made for the proper cost considerations, but they also cause misallocation because of their failure to reflect demand pressures.

Ironically, one of the biggest shortcomings of the Soviet pricing system is what the Soviets consider to be one of its important advantages – its immunity to outside fluctuations. This isolation makes it possible for the Soviets to maintain relatively fixed prices regardless of what is going on in the capitalist world. This in turn ensures relatively stable prices, which encourage Soviet managers to make long-range plans without having to worry about confusing and distorting price adjustments. Such a system, however, assumes that there will be no fundamental change in demand and supply relationships similar to the one we have seen in the mid-1970s. It is conceivable that, since the USSR is isolated and relatively unaffected by the world economic system, the radical changes that have taken place in the aftermath of the energy crisis have had little effect on the Soviet economy. After all, the ratio of exports as a percentage of GNP is less than 4 per cent in the Soviet Union, one of the lowest percentages in the world. True, world price fluctuations are of little consequence for most of the Soviet economy, but that does not apply to petroleum and natural gas. World fluctuations do affect the liquid fuel industry in the Soviet Union, because not only is it the world's largest producer of petroleum, but it is also the world's second largest exporter after Saudi Arabia. As mentioned earlier, it has exported between 25 and 30 per cent of its total petroleum production during the last few years. In the case of natural gas, the Soviet Union is the world's second largest producer and the third largest exporter. On a gross basis, it exports about 10 per cent of its gas production.

Given the relative importance of world trade to the Soviet petroleum, coal, and gas industries, it was economically irrational for Soviet price-makers to ignore the fourfold increase in petroleum prices that took place in 1973. Yet in spite of what has been happening in the rest of the world, 1967 marked the only major change in the wholesale price of Soviet coal, oil, and natural gas during the last decade and a half. Then, in what at the time was viewed as a dramatic step, the Soviets increased the wholesale price of all raw materials an average of 73 per cent.[63] But since then there have been only minor adjustments. Clearly, as the world petroleum price continues to increase, the distortion between Soviet and world prices also increases. Irrational as it may sound, wholesale fuel prices in the USSR in 1977 were actually slightly lower than they were in 1970.[64] Inevitably this divergence is liable to set off a new set of distortions and economic waste.

With the Soviet wholesale price virtually the same as it was in the late 1960s, it should come as no surprise to learn that the failure to increase prices has had an adverse effect on the Soviet Ministry of the Petroleum Industry's (MPI's) profitability rate. Thus, the MPI's profitability rate, which was as low as 5·7 per cent in 1965, rose to 27·8 per cent in 1970 in the aftermath of the 1967 wholesale price hike, but fell again as Soviet petroleum officials had to turn more and more to the Far East and the Arctic region for their petroleum. Inevitably, with soaring costs and frozen wholesale prices, there was only one direction in which profits could go. As measured by official internal Soviet data, the rate of profitability in 1977 was back to 13·2 per cent.[65] Similarly, the profit rate for gas, which was 9·3 per cent in 1965, rose to 64·5 per cent in 1970, but also fell to 19·9 per cent in 1977. For the same reason, the rate of profit in the Coal Ministry fell from 6 per cent in 1970 to a deficit of 0·8 per cent in 1977 and required an operating subsidy.[66] The peat industry suffered a similar drop and operated at a loss of 0·6 per cent. Profitability for all energy industries was down to 10 per cent.

Obviously, if wholesale prices of energy in the Soviet Union were allowed to reflect world prices, the rate of profit of the various production ministries would be much higher, but then Soviet planners would be tempted to divert all of their output to foreign markets. This would cause chaos in the domestic planning system. Yet, if prices were higher, Gosplan, the Soviet planning agency, might be more willing to increase the amount of investment allocated for the development of new but seemingly unprofitable oil fields. As matters now stand, the planners are getting mixed signals. Domestic petroleum prices and profit ratios tell them they should divert investment resources elsewhere. Foreign prices and profit ratios tell them the opposite. Any shadow prices they adopt to provide them with a more realistic but unofficial view of costs and opportunities will have to be very carefully designed to head off what otherwise could be a disaster in the way Soviet investment resources are allocated. As we shall see, until 1977 confusion negatively affected efforts to increase investment efficiency.

THE CHANGE IN THE FUEL BALANCE

In light of all the shortcomings of the Soviet pricing system and their deleterious impact on the efficient use of raw materials, it would appear that on balance capitalist prices are doing a better job in guiding us as to how we should run our economy. However, that does not mean that this is always the case, particularly when it comes to dealing with non-renewable resources. With time, we have become considerably more humble than we once were. For example, Western economists for many years derided Soviet planners for sticking with coal so long after virtually everyone else in the world had switched to oil and gas.[67] Much of this backwardness was attributed to the unresponsiveness of the Soviet price system, and its failure to reflect the advantage that increased use of petroleum would bring. While presumably Soviet planners could have ordered any planning switch that they wanted, regardless of what the comparative costs of using oil or coal might have indicated, many observers believed that at least some of the reluctance to make this shift was due to the failure of Soviet planners and managers to perceive not only how much cleaner, but how much 'cheaper', oil and gas were. In any case, the Soviets held off on such a decision until the mid-1950s. Then, as Campbell notes, they had 'a radical reorientation ... in their perception of the relative cost of fuel' and began in earnest to switch away from coal, lignite, and peat to oil and gas.

In retrospect, the Soviets were probably lucky that the switch came so late, because the longer it is since an economy has switched away from coal, the harder it is to switch back. Of course, if the Soviets had not made the shift to oil, they probably would have denied themselves some important efficiencies and economic growth benefits that petroleum made possible, at least in the short run. They also probably lost some growth from shifting as late as they did. In the long run, however, economists may some day say that the Soviets, like everyone else, would have been better advised to have hung on longer to the coal-dominated economy. In that event, the Soviet price system with all its foibles may come to be praised for its part in delaying the shift as long as it did.

The price system notwithstanding, the replacement of solid fuels with liquid fuels began in earnest in the late 1950s. It should not be forgotten, however, that no matter how much of a stimulus or deterrent the price system may have been towards making such a change, the shift would not have occurred if there had not been the discovery of the Volga-Ural fields and subsequently the west Siberian deposits. At the time foreign trade was even less important than it is now, and it is most unlikely that the Soviets would have resorted to importing such large quantities of petroleum. Thus the switch was as dependent on the availability of such resources as on any incentive scheme involving an alignment of price or any reordering of physical planning targets.

GAS

Until the Second World War, most of the gas produced was either gas associated with the production of petroleum as a by-product or gas derived from the gasification of coal.[68] Firewood was often a more important source of fuel than gas. (See Table 3.1.) The realization that gas had valuable economic uses and the discovery of major gas fields led to some important changes after the Second World War.[69] During the war large fields were found in the Volga-Urals, the Caucasus, and Shebelinka in the central Ukraine. Only after that, in the mid-1950s, did the output of gas from gas fields grow to exceed gas derived as a by-product from the lifting of petroleum.[70] The discovery in the Ukraine was followed by other discoveries in Uzbekistan, Turkmenistan, and the giant fields at Orenburg and in west Siberia at Urengoi, Medvezhze, and Zapoliarnoe. By 1979, the Soviet Union was thought to hold 39·9 per cent of the world's proven reserves.[71] Iran ranked second to the Soviet Union, with 16 per cent. As these new fields were exploited, associated gas from the petroleum industry became a smaller percentage of total production. By the mid-1950s, it made up less than one-half of the total production. In 1977, it constituted 17 per cent of total production.[72] As oil production increased, particularly in remote areas, the Soviet planners, as we saw, simply did not provide for energy processing or gas pipeline facilities, so that the amount of associated gas that was processed fell from 70 per cent in 1965 to 59 per cent in 1975.[73] The rest was wasted.

Table 3.1   *Composition of Fuel Used in Boilers and Furnaces* (%)

	1970	1975	1980 (Plan)
Coal	37·9	32·9	29·5
Coke	7·6	6·6	5·9
Oil	15·5	18·3	14·7
Natural Gas	25·4	29·6	34·8
Peat	1·7	1·2	1·2
Shale	0·7	0·7	0·6
Wood	2·0	1·4	0·9
Liquified Gas	0·8	0·9	1·1
Coke Oven Gas	1·9	1·7	1·5
Blast Furnace Gas	2·5	2·2	1·9

Source:
Nekrasov and Pervukhin, p. 149.

When the potential of this gas was finally realized, a network of pipelines was built, so that gas eventually became a major source of energy in Soviet cities and an important export commodity to Eastern and Western Europe. It also became increasingly important in the overall production of Soviet energy. (See Table 3.2.) The flow of gas to the cities for both industrial and household uses turned out to be one of the most important factors in reducing atmospheric pollution in Soviet cities. The burning of coal in urban areas was sharply curtailed. In many cases it was a most welcome relief, particularly because the level of sulphur dioxide and fall-out had caused serious health problems.[74]

Table 3.2   *Consumption of Energy in the USA and USSR by Major Energy Source* (%)

| | USA | | | USSR | | |
	Coal	Oil	Gas	Coal	Oil	Gas
1940	52	31	11	75	23	2
1945	51	30	13			
1946				78	19	3
1950	38	40	18	77	20	3
1955	29	44	23	75	23	3
1960	23	45	28	62	29	9
1965	22	44	30	50	32	18
1970	19	44	33	41	36	23
1971	18	45	33	39	36	25
1972	17	46	32	38	38	24
1973	18	47	30	37	39	24
1974	18	46	30	36	39	24
1975	18	46	28	35	40	25
1976	19	47	27	34	39	27
1977	19	49	26			

Sources:
(1) United States. US Bureau of the Census, *Statistical Abstract for the United States: 1977*, 8th edn (Washington, DC, 1977), p. 594; ... *1978*, p. 604.
(2) Soviet Union. A. M. Nekrasov and M. G. Pervukhin, *Energetika SSSR v 1976–1980 godakh* (Moscow: Energia 1977), p. 146; Tsentral'noe statistichesko upravlenie, *Narodnoe khoziaistvo SSSR v 1959* (hereafter *Nar. khoz.*) (Moscow: Statistika, 1959), p. 176; *Nar. khoz.*, 1970, p. 183; *Nar. khoz.*, 1976, p. 204; Ministerstvo Vneshnei Torgovli, *Vneshniaia torgovlia SSSR v 1976 gody* (Moscow: Statistika, 1977), and for earlier years.

As mentioned earlier, the switch from coal to gas and oil came much earlier in the United States. Whereas in the 1940s the United States, like almost every other country in the world, relied on coal for over 50 per cent of its energy, by 1950 coal's share in the total US energy balance had dropped to under 40 per cent. (See Table 3.2.) Petroleum and natural gas in the United States so supplemented coal that there was a fall not only in the relative share of coal consumed but also in the absolute amount consumed. The quantity consumed in tons increased again only in the late 1960s, but until very recently the relative share of coal in the United States has continued to diminish. Offsetting that, oil's share has remained consistently over 40 per cent since the 1950s. By

1976, it accounted for approximately 47 per cent of all energy consumed in the United States. The growth in the use of natural gas in the United States was even faster, at least until about 1972. Whereas gas made up only 11 per cent of all energy consumed in 1940, by 1960 it supplied more energy than coal. In 1971, at its peak, it was the source of 33 per cent of all the energy generated in the United States.[75]

As we saw, coal remained the main source of energy in the Soviet Union for a much longer period of time. In 1940, Soviet coal accounted for about 75 per cent of all Soviet energy consumed. (See Table 3.2.) Reflecting the different trends in the two countries, the relative swing away from coal in the Soviet Union began only in 1955. By then, even though coal constituted less than 30 per cent of America's energy, it still was the source of over 70 per cent of all energy consumed in the USSR. It was not until 1966 that coal's relative share was less than 50 per cent. Thereafter, coal's relative importance fell rapidly, so that by 1976, as the relative importance of oil and gas rose sharply, coal accounted for less than 30 per cent. None the less, the consumption of natural gas in the Soviet Union has not come to exceed coal as happened in the United States, and even now Soviet domestic consumption of oil is not much larger than coal. (See Table 3.2.) Of the three fuels, oil made up 39 per cent of Soviet energy consumption in 1976 and coal 34 per cent. The contrast with the United States and its heavy dependence on petroleum is striking.

When analyzing the Soviet fuel balance and determining how dependent the Soviet Union has become on petroleum, Western specialists sometimes confuse Soviet production with consumption.[76] This is because the Soviets usually publish only the relative production statistics.[77] (See Table 3.3.) By relying only on the production figures, Western observers sometimes conclude that the relative use of petroleum in the Soviet Union approximates that in the United States. Since the relative share of Soviet petroleum in total energy production in

Table 3.3   *Fuel Balance – Production of all Energy Forms* (%)

	1975	1980 (Plan)
Oil and Gas Condensate	43·0	43·1
Gas Natural and By-Product	21·2	24·5
Coal	30·0	26·0
Peat	1·1	0·9
Shale	0·7	0·5
Hydro-electric	2·6	3·0
Nuclear	0·4	1·4
Wood	1·0	0·6

Source:
Nekrasov and Pervukhin, p. 149.

1976 was 45 per cent, it does indeed approximate the American figure for consumption.[78] What is often ignored is the fact that the Soviets are net exporters of 25 to 30 per cent of their oil production but only 3 per cent of their coal and 4 per cent of their gas. This means that the percentage of petroleum in the consumption fuel balance will be much smaller than the same percentage in the production fuel balance. Those who neglect to make the distinction may be misled into believing that Soviet reliance on petroleum for domestic consumption of energy is much greater than it actually is.

CONCLUSION

In the post-Second World War era the Soviet Union opened up vast new oil and natural gas fields. Soviet efforts were hampered, however, by the inappropriateness on occasion of their technology and their planning and incentive system. As a result, there was often enormous waste. Nor were their efforts eased by the remoteness of some of their large discoveries. Yet, despite the sometimes very large costs involved and the need to call for foreign help, the Soviet Union at least until the late 1970s has managed reasonably well. Its output of oil and gas has increased annually, so that it is able to satisfy not only its domestic needs, but also an increasingly large portion of its hard-currency export requirements as well. However, when such a large quantity of petroleum was suddenly offered on the world market, it was bound to be disruptive. It is to the export of Soviet petroleum and its impact on world markets prior to 1973 that we turn now.

NOTES: CHAPTER 3

1   *Pravda*, 10 December 1963, p. 1; *VT SSSR*, 1913–1940, pp. 58, 144.
2   *VT SSSR*, 1971, pp. 32, 46.
3   See Table 2.3; Elliot, p. 74; Sutton, Vol. II, p. 90.
4   Sutton, Vol. II, p. 89; Antony C. Sutton, *Western Technology and Soviet Development, Vol. III, 1945 to 1965* (Stanford: Hoover Institution Press, 1973), pp. 38–9, 135–6.
5   Campbell, 1968, p. 126.
6   ibid., pp. 127–8.
7   ibid., p. 125; Robert W. Campbell, *Trends in the Soviet Oil and Gas Industry* (Baltimore: Johns Hopkins University Press, 1976), p. 27.
8   CIA Oil, July 1977, p. 15.
9   Elliot, p. 96; 'Europe-Sibir'', *Ekonomika i organizatsiia promyshlennogo proizvodstva* (hereafter *EKO*) no. 4, 1976, pp. 160–7.
10 *Total Information*, no. 68, 1976, Geneva, Switzerland, p. 2.
11 *Total Information*, p. 4; Elliot, p. 97; *The Review of Sino-Soviet Oil*, April 1979, p. 50.
12 Sutton, Vol. I, p. 24; Tolf, pp. 65, 182.

13 Campbell, 1968, pp. 102–20, 126–8; Joseph S. Berliner, *The Innovation Decision in Soviet Industry* (Cambridge: MIT Press, 1976).
14 Campbell, 1968, p. 93.
15 ibid., p. 109.
16 ibid., p. 93.
17 *Pravda*, 28 February 1978, p. 2.
18 *Current Digest of the Soviet Press*, 14 February 1973, p. 6.
19 *Pravda*, 20 November 1962, p. 4.
20 *Review of Sino-Soviet Oil*, November 1977, p. 10.
21 Campbell, 1968, p. 108.
22 ibid., p. 109.
23 ibid., p. 111.
24 Campbell, 1976, p. 20.
25 Personal communication with the American manufacturer.
26 Campbell, 1968, p. 115.
27 CIA Oil, July 1977, p. 26.
28 Personal communication with American manufacturers.
29 Campbell, 1968, p. 103.
30 CIA Soviet, April 1977, p. 5.
31 *Review of Sino-Soviet Oil*, January 1978, p. 26; 17 April 1979, pp. 10–11; *Neftiannoe Khoziaistvo*, 6 June 1968, p. 4.
32 CIA Oil, July 1977, p. 26.
33 ibid., p. 25.
34 *Oil and Gas Journal*, 18 September 1978, p. 66.
35 Elliot, p. 78.
36 Campbell, 1968, p. 107.
37 'Otsenka prirodnykh resursov', *Voprosy geografii* (Moscow: Mysl', 1968), p. 47.
38 T. Khachaturov, 'Prirodyne resursy i planirovanie narodnogo khoziaistva', *Voprosy ekonomiki*, August 1973, p. 17.
39 Marshall I. Goldman, *The Spoils of Progress* (Cambridge: MIT Press, 1972), p. 49.
40 N. K. Feitel'man, 'Ob ekonomicheskom otsenke mineral'nykh resursov', *Voprosy ekonomiki*, November 1968, p. 110; Iu. Iakovetz, 'Dvizhenie tsen mineral'nogo syr'ia', *Voprosy ekonomiki*, June 1975, p. 3; S. Levin, V. Vasil'eva and N. Kosinov, 'Tsenoobrazovanie v neftianoi promyshlennosti', *Planovoe khoziaistvo*, July 1976, p. 111.
41 *Trud*, 12 August 1967, p. 2.
42 *Literaturnaia gazeta*, no. 7, 12 February 1975, p. 10.
43 Khachaturov, pp. 20–1.
44 ibid., p. 26; *Sotsialisticheskaia industriia*, 3 March 1978, p. 2, 31 May 1978, p. 2.
45 Shell Briefing Service, *Enhanced Oil Recovery* (London, Shell Oil Company, March 1979), p. 4.
46 K. E. Gabyshev, 'Ekonomicheskaia otsenka prirodnykh resursov i rentnye platezhi', *Vestnik Moskovskogo universiteta, seriia ekonomika*, no. 5, 1969, p. 17; Robert Stobaugh and Daniel Yergin (eds), *Energy Future* (New York: Random House, 1979), p. 87.
47 ibid., p. 18; *Sotsialisticheskaia industriia*, 8 January 1971, p. 2.
48 Khachaturov, p. 17; Martsinkevich, p. 65.
49 N. G. Feitel'man, 'Sotsial'no-ekonomicheskie problemy ekologicheskogo ravnovesiia v zapadnoi Sibiri', *Voprosy ekonomiki*, October 1978, p. 11; T. Khachaturov, 'Ekonomicheskie problemy ekologii', *Voprosy ekonomiki*, June

　　　　1978, p. 6. This was also reported to me in a personal conversation in Moscow in
　　　　December 1978.
50　*Sotsialisticheskaia industriia*, 12 August 1978, p. 2.
51　ibid.; Feitel'man, 1978, p. 1.
52　CIA Oil, July 1977, pp. 12, 14.
53　Feitel'man, 1978, p. 1.
54　T. Khachaturov, 'Ekonomicheskie problemy ekologii', *Voprosy ekonomiki*, June
　　　　1978, p. 6.
55　G. Mirlin, 'Effektivnost' ispol'zovaniia mineral'nykh resursov', *Planovoe
　　　　khoziaistvo*, no. 6, 1973, p. 32; *Review of Sino-Soviet Oil*, May 1976, p. 23.
56　Central Intelligence Agency, *The USSR: Development of the Gas Industry*, ER
　　　　78-10393 (Washington, DC, July 1978), p. 48 (hereafter CIA Gas, July 1978);
　　　　Current Digest of the Soviet Press, 8 February 1978, p. 2.
57　*Pravda*, 27 January 1978, p. 2; Daniel Park, *Oil and Gas in COMECON
　　　　Countries* (London: Kogan Page, 1979), p. 67.
58　*Turkmenskaia iskra*, 6 December 1977, p. 2; *Literaturnaia gazeta*, 18 January
　　　　1978, p. 10.
59　Khachaturov, 1973, p. 27.
60　ibid., p. 27; *Trud*, 12 August 1967, p. 2; *Sotsialisticheskaia industriia*, 24
　　　　March 1978, p. 1; *Current Digest of the Soviet Press*, 5 November 1977, p. 1.
61　'Berech' i umnozhat' prirodnye bogatstva', *Planovoe khoziaistvo*, no. 6, June
　　　　1973, p. 5; *Pravda*, 11 September 1978, p. 2.
62　CIA Gas, July 1978, p. 4.
63　Ia. Kovetz, p. 4.
64　Tsentral'noi Statisticheskoe Upravlenie, *Narodnoe khoziaistvo SSSR v 1977
　　　　gody* (Moscow: Gostatizdat, 1978), p. 142 (hereafter *Nar. khoz.* and the
　　　　appropriate year); *Nar. khoz.*, 1974, p. 212.
65　These figures were provided to me by a Soviet economist in Moscow in late 1977.
　　　　See also *Nar. khoz.*, 1977, p. 544.
66　*Current Digest of the Soviet Press*, 7 March 1979, p. 11.
67　Campbell, 1968, pp. 9, 10, 178.
68　Jonathan P. Stern, *Soviet National Gas Development to 1990* (Lexington,
　　　　Mass.: Lexington Books, 1980), p. 21.
69　CIA Gas, July 1978, p. 2.
70　ibid., p. 1.
71　Stern, p. 39; *Petroleum Economist*, August 1979, p. 313.
72　CIA Gas, July 1978, pp. 1–2.
73　Jack H. Ray, 'The role of Soviet natural gas in East-West cooperation', Vienna,
　　　　mimeo. prepared for Vienna II Conference, 5–8 March 1979, p. 32.
74　Marshall I. Goldman, *The Spoils of Progress* (Cambridge: MIT Press, 1972),
　　　　p. 127.
75　US Bureau of the Census, *Statistical Abstract for the United States; 1977*, 98th
　　　　edn (Washington, DC, 1977), p. 594.
76　Elliot, p. 7; *Review of Sino-Soviet Oil*, January 1976, p. 21; *Soviet Power
　　　　Reactors*, 1974, Report of the United States Nuclear Power Delegation Visit to
　　　　the USSR, 19 September to 1 October 1974, USERDA, ERDA-2 (Washington,
　　　　DC, 1974), p. 10; Zeman and Zoubek, p. 64; Park, p. 43.
77　*Nar. khoz.*, 1977, p. 204.
78　This is calculated in terms of million tons of standard fuel equivalent – standard
　　　　fuel is equal to the heat value of 7,000 kilocalories per kilogram.

4

The Post-1939 Period to 1973 – External Developments

With its oil fields in a state of postwar disrepair and confusion, the Soviet Union was hard pressed to satisfy its domestic energy needs. In the circumstances, the Soviets were more interested in imports than exports and they were not always subtle as to how they obtained them. But whether supplies were scarce or in surplus, the Soviets have always used oil as a major weapon of economic and political diplomacy. During the subsequent years of oil abundance that marked the period from the early 1950s to the Yom Kippur War, oil was used by the Soviet Union to reward friends, penalize enemies, and tempt the innocent and exploited. Contracts and commercial niceties were never allowed to stand in the way of political priorities. In other words, the Soviets behaved just as they had in the 1930s, as they were to again in the 1970s, and as almost everybody else in the capitalist world would have behaved and did behave.

The Soviet Union's economic attitude towards the newly liberated countries of Eastern Europe after the Second World War was ambiguous. Even though these countries were all to become communist satellites and allies, initially the Soviet Union dealt harshly with several of them, particularly those which had participated as members of the German–Italian Axis. In those instances, the Soviets exacted stiff reparation payments. Thus it was standard procedure for the Soviets to dismantle and take home many of the more desirable factories that had survived the war. One estimate is that overall they removed over $10 billion worth of equipment.[1] In other instances their demands took the form of a claim to joint ownership in the East European factories, shipping firms, or airlines.[2]

This policy was of particular consequence for Rumania. The Soviets dismantled some 700 Rumanian factories, including oil drilling and refinery equipment or facilities. In addition, the Soviets made themselves equal partners in sixteen Rumanian–Soviet joint stock companies. The first and most important of these joint stock companies was an oil company, Sovromneft, that was founded in July 1945.

Subsequently the Soviets also created Sovromgaz and Sovrommutilaj (oil equipment) during the period from 1948 to 1952.

These joint ventures were quite important to the Soviets. Not only did they share in profits, but they also arranged for the products of these companies to be sent to the USSR at unusually favorable prices. In the case of Sovromneft, this meant petroleum products were exported at 1938 prices. Considering that the Soviets contributed no capital to the operation of such ventures (their share consisted solely of confiscated German assets that were already in the country), this was a very profitable operation for them. From the founding of Sovromneft until its dissolution, the Soviets imported over 21 million tons of refined petroleum products from Rumania.[3] Given the fact that Soviet refineries were in a state of disrepair for a portion of that time, these imports were of vital importance. Indeed, the Soviet Union was a net importer of crude and refined products until 1954, and approximately 80 to 90 per cent of it came from Rumania. Many of the joint companies were broken up after the East European riots in 1953, although the Soviets held on to Sovromneft until December 1955.

While initially the Soviets had to worry more about imports than exports, it none the less resumed exports immediately after the war's end. However, the Second World War and the Communist takeovers in Eastern Europe marked a radical reorganization of Soviet export policy. Soviet trade had fallen off sharply after the early 1930s, but prior to the war what little the Soviets did buy and sell went to a wide variety of countries around the world. With the coming of communism to the countries of Eastern Europe, the Soviets all but halted their sales and purchases to countries outside the new communist bloc.[4] Whereas the Soviet exports in 1930 to pre-communist China and to what was to become the Council of Mutual Economic Assistance bloc (CMEA) amounted to 20 per cent of the total; by 1950, when these countries had become communist, the total was closer to 80 per cent.

The flow of petroleum reflects this reorientation. As early as 1946, Bulgaria, Poland, Czechoslovakia, and Yugoslavia had become the largest importers of Soviet petroleum products. They remained the main importers of Soviet petroleum products into the mid-1950s, although there were some significant changes. After the takeover by the Chinese Communist Party in 1949, Chinese imports doubled and continued to grow. By 1959, China was importing over 2 million tons of Soviet petroleum a year. In fact, from 1955 to 1961 China was the Soviet Union's largest communist customer by a significant margin. At its peak in 1955, the Soviet Union was exporting as much as 24 per cent of all its refined petroleum products to China.[5]

However, just as politics seemed to dictate who amongst the communist countries would be sent Soviet oil, so politics determined who would be denied it. The Yugoslavs have hinted that the Soviets forced Sovromneft to suspend oil shipments in April 1948 as a way of making Tito conform to Soviet dictates. Petroleum shipments from the USSR itself to Yugoslavia were halved in 1948 from the year earlier and then reduced to a mere 15,000 tons in 1949, after which they ceased completely until 1954. Similarly, after the Soviets decided to protest about the Israeli invasion of Sinai in 1956, Siouznefteeksport broke its contract and suspended petroleum shipments, which had reached a level of about half a million tons a year.[6]

Oil was also used as a weapon in Soviet dealings with China. Initially it flowed freely, and the Soviets helped to look for oil when they created a Chinese–Soviet joint company – Sovkitneft – in 1950. But as with Yugoslavia, once political relations deteriorated, so did the export flow. The Soviet technicians sent to help Chinese industry were called home in mid-1960. Oddly enough, Soviet exports of petroleum continued at a relatively high level until 1963 and 1964. But then, when the split between the Soviet Union and China, which until then had been relatively muted, became vehement, Soviet exports, which had been 1·4 million tons in 1963 and 505,000 tons in 1964, shrivelled to about 40,000 tons in 1965 and 1966 and eventually ceased entirely.

So far, nothing has been said about Hungary and East Germany. Since they had their own refining capacity, Russian refined petroleum products were not of much use to them. Instead, as Soviet petroleum output began to recover, Hungary and East Germany became the largest purchasers of Soviet crude oil.[7] Apparently some of this crude oil was then refined by the East Germans and sent back to the Soviet Union. Next to Rumania, East Germany was the other large supplier of refined products to the Soviet Union. Ultimately, during the early 1950s, all of the other East European countries except for Bulgaria and Rumania also increased their purchases of Soviet crude oil.

Except for 1948, Soviet petroleum exports rose every year until 1978. (See Table 2.3.) But since exports in the immediate postwar period were at a very low level and did not reach 10 million tons until 1956, that sounds more impressive than it was. Also it took until 1954 before Soviet petroleum exports exceeded imports. It was possible for the East Europeans to make do with such a low level of imports, in part because they had few automobiles, and more important, like the USSR, because they still had primarily coal-oriented economies. But like the Soviet Union, belatedly, they too began to de-emphasize coal as a source of fuel for the same reasons, in order to reduce pollution and increase economic efficiency. However, this was impossible to implement without petroleum imports, since, except for Rumania, the other members of CMEA had virtually no petroleum of their own. Of the other countries in Eastern Europe, only Hungary produced more than 1 million tons of crude oil a year, and even for Hungary its output represented only about 22 per cent of its total consumption.[8] The main source of supply for Poland, the German Democratic Republic, Bulgaria, Hungary, and Czechoslovakia was the Soviet Union. The Soviet Union typically has supplied 80 per cent and more recently 90 per cent of all East European petroleum imports. For a country like Czechoslovakia, the Soviet Union supplied about 95 per cent of its total petroleum consumption, and in some instances 100 per cent.[9]

For many years the Soviets seemed quite pleased with such an arrangement. It provided them with an important export product for their East European customers as well as 10 per cent of their total export earnings. To many outsiders, this began to look like a form of economic imperialism. As they saw it, the East European countries were being drawn into an economic web with their factories tied to Soviet raw materials. This dependence was increased with the completion of the Druzba petroleum pipeline in November 1964 from the Soviet Union to Poland, East Germany, Hungary, and Czechoslovakia.[10] In the same way, the opening of the Bratstvo (fraternal pipeline) in 1967 to Bratislava in Czechoslovakia tied the Czechs, the Poles, and eventually the East Germans, Hungarians, and Bulgarians to Soviet gas.[11]

However, it became increasingly evident that not all of the East Europeans were happy with these arrangements. Complaints reached the West in the wake of Tito's disaffection and the subsequent trials of Slansky in Czechoslovakia and the vice-premier of Bulgaria, and the other would-be Titos, in the late 1940s. The most common accusation was that they were being exploited by the Soviets who appeared to be charging higher prices to Eastern Europeans than to others. These complaints continued off and on into the 1950s and 1960s. In a series of studies written in 1959 and the 1960s, a Michigan University economist Horst Mendershausen seemed to show just that.[12] However, in a challenge to Mendershausen, Franklyn Holzman pointed out that this apparent discrimination was actually the result of the artificial nature of the East European trading bloc. It was true that the Soviet Union discriminated against the East Europeans, but the East Europeans reciprocated by also charging the Soviet Union higher prices for their exports than they charged other non-Communist countries.[13] The result was mutual discrimination. According to Holzman, this was inevitable, given the artificial 'customs union' effect of the communist trading arrangement. The Soviets charged more for their exports to Eastern Europe, but balanced this off by paying more for their East European imports.

The theoretical elegance of the argument failed to persuade most of the East Europeans that they had nothing to complain about. All they knew was that in 1960 while the Soviets were charging the East Germans 18 rubles a ton for crude oil, they were charging the West Germans and Italians only 9 rubles a ton.[14]

It was somewhat surprising, therefore, when in the mid-1960s some Soviet economists did indeed begin to insist that they were not the exploiters but the exploited.[15] As they saw it, the Soviet Union was not the colonizer, but the colonized. While the Soviet Union was busy supplying raw materials to its East European allies, its allies were using the Soviet Union as a dumping ground for their manufactured goods. In order to supply these raw materials, the Soviet Union was being forced to turn more and more to the East and Siberia to supply its own needs. This meant not only increased labor costs, but more pressure on Soviet capital and transportation. Thus, while there may have been short-run merit in making the East Europeans dependent on the Soviet Union for their raw materials, with time it became clear that there were also long-run disadvantages to such a policy. For this reason, Soviet economists began to urge that the East European colonies seek additional sources of supply elsewhere. In what clearly was a major change of emphasis, in 1965, the East Europeans were told to redesign their foreign aid program, so that they would be compensated by the developing country recipients with oil and ferrous and non-ferrous metals.[16] As one of them put it,

> the steadily growing demand of the CMEA member countries for oil and the desire of these countries to improve their consumption can be met not only by deliveries of oil via the Druzba [Friendship Oil Pipeline] and expansion of their own oil production, but also from the developing countries of the Middle East and the African continent. Increased oil deliveries from these countries can be carried out at the cost of rendering technical assistance to the developing countries in establishing their national oil industries and by engaging in other forms of cooperation.[17]

In other words, they had better become less dependent on the Soviet Union. This theme was to be repeated continually in the years to come.[18] At the same time, if the East Europeans expected to continue importing raw materials from the Soviet Union, they had better bear some of the increased capital costs of exploration, development, and transportation.

One of the best examples of the joint investment effort that followed was the Orenburg–Soiuz pipeline. Perhaps the most expensive and encompassing of all such projects undertaken, the Orenburg–Soiuz pipeline necessitated the construction of a 2,750-kilometer (1,700-mile) pipeline at a cost of approximately 2·4 billion rubles.[19] The cooperative aspects of the project involved the assigning of responsibility for construction of a specific portion of the pipeline to each of the ultimate recipients. Under the plan, Bulgaria agreed to join with the Soviet Union to construct the 558 km segment from Orenburg to Alekhsandrov-gay.[20] From there Czechoslovakia was to build the 562 km segment to Sokhranovka. The Poles extended it then to Kremenchug from 596 km where it was taken up by the East Germans to Bar for another 518 km. The final segment of 516 km to the Soviet border was to be built by the Hungarians. In return for these efforts, each of the five countries was to receive 2·8 billion cubic meters of natural gas a year. Rumania, which was not involved in the construction of the pipeline, was assigned instead to finance the French construction of a sulphur removal facility at Orenburg. For that it will draw 1·5 billion cubic meters off the pipeline. The Soviet Union's total annual commitments to Eastern Europe under the Orenburg Agreement amounted to 15·5 billion cubic meters a year.

Considering the Soviet's shortage of pipe-making capacity, the project seems like a sensible one. While the Orenburg field was discovered as early as October 1966, the Soviets seemed unable to undertake the Orenburg–Soiuz pipeline in addition to all their other ongoing efforts.[21] They began work on a pipeline in 1971, but brought the CMEA countries in on June 1974, in order to open up additional processing and shipping facilities.[22] It turned out, however, that not all of the East European parties had the know-how or manpower to complete their segments. First, most of them had to purchase Western pipe and the 158 turbines that went into equipping the 28 compressor pumping stations.[23] Secondly, not all of them had the work crews with the experience needed to complete such a massive undertaking. Originally it was estimated that a workforce of 30,000 from all six countries would be needed. But while the East Germans managed to complete about half of their work and the Poles all of it, countries like Hungary, Czechoslovakia, and Bulgaria were able to fulfill hardly any of their pipelaying assignments.[24]

Despite some delays and the inevitable problems of co-ordinating such a massive effort, the project was completed on 29 September 1978.[25] The contract calls for the delivery of 8 to 9 billion cubic meters of gas in 1979 and a total commitment of 15·5 billion cubic meters in 1980. For the East Europeans, that will mean a substantial increase in the amount of Soviet natural gas they import. Orenburg will mean an increase of approximately 90 per cent for the Bulgarians, about 60 per cent for the Czechs, 80 per cent for the East Germans, 150 per cent for the Hungarians, and about 100 per cent for the Poles. (See Table 4·1.) Spread over twenty years, each country except Rumania will import about 55 billion cubic meters, which is roughly equivalent to 48 million tons of oil, or 2·4 million tons of petroleum, a year.[26] Furthermore, the

Table 4.1 Natural Gas Imports and Exports of the USSR (million cubic meters)

	1965	1966	1967	1968	1969	1970	1971	1972	1973	1974	1975	1976	1977
Imports													
Afghanistan			207	1,500	2,030	2,591	2,513	2,849	2,735	2,847	2,853	2,500	2,400
Iran						965	5,623	8,197	8,680	9,094	9,559	9,300	9,400
Total			207	1,500	2,030	3,556	8,136	11,046	11,415	11,941	12,412	11,800	11,800
Net Imports						256	3,581	5,976	4,583				
Exports													
Bulgaria										307	1,185	2,200	3,000
Czechoslovakia			265	587	889	1,342	1,639	1,937	2,363	3,231	3,694	4,300	4,700–4,900
E. Germany									760	2,900	3,302	3,400	3,600
Hungary									24		601	1,000	1,000
Poland		828	1,025	1,000	994	1,002	1,488	1,500	1,710	2,117	2,509	2,500	2,800
Austria				142	782	956	1,428	1,633	1,622	2,106	1,883	2,800	2,800
France											5	1,000	2,000
W. Germany									353	2,145	3,097	4,000	5,800
Italy										790	2,342	3,700	5,500–6,000
Finland										443	719	900	800
Total	392	828	1,290	1,729	2,665	3,300	4,555	5,070	6,832	14,039	19,337	25,800	32–33,000
Net Exports	392	828	1,083	229	635					2,098	6,916	14,000	20–21,000

Source:
Various issues of *VT SSSR*.

Table 4.2 USSR Production of Crude Oil and Export of Oil and Oil Products by Country

	million tons						million rubles***					
	1972	1973	1974	1975	1976	1977**	1972	1973	1974	1975	1976	1977
Production												
Crude oil	400	429	459	491	520	546						
Exports												
Austria	1·0	1·3	1·0	1·3	1·5	2·0	14·9	28·6	61·3	77·9	99·1	147·7
Belgium	2·5	1·7	1·8	1·3	2·1	2·0	35·7	68·6	116·2	79·1	139·0	146·3
Canada			0·2	0·2	0·1	0·3			8·0	11·7	6·0	23·0
Denmark	0·8	0·6	0·7	1·2	1·6	2·1	10·2	32·5	43·1	67·8	108·7	160·0
France	3·1	5·3	1·4	3·3	5·7	5·4	43·3	91·0	84·2	192·6	372·3	390·6
Germany*	6·6	6·3	6·9	8·4	8·2	8·9	94·7	236·3	463·8	535·3	656·4	783·2
Greece	0·9	0·8	1·0	1·9	1·9	2·7	17·2	17·6	69·2	112·4	130·5	207·6
Iceland	0·4	0·5	0·5	0·4	0·4	0·4	8·6	12·9	34·6	32·9	32·3	37·6
Italy	8·4	8·7	6·8	6·9	12·0	10·5	109·0	153·1	394·9	392·3	783·4	744·3
Japan	1·0	2·0	1·2	1·3	1·8	0·9	15·6	41·2	71·4	67·6	113·1	62·6
Netherlands	2·4	3·2	3·0	3·1	2·7	3·0	40·2	135·6	202·0	201·0	220·6	268·3
Norway	0·4	0·6	0·3	0·3	0·2	0·7	6·7	11·4	20·9	18·0	15·0	54·5
Spain	0·8	0·5	1·4	1·7	2·0	1·7	11·5	9·1	81·7	100·3	127·8	120·7
Sweden	4·4	3·2	3·0	3·5	2·7	2·8	58·0	58·3	168·0	172·8	167·0	192·0
Switzerland	0·8	0·7	0·8	1·0	0·9	0·9	14·0	36·1	52·8	61·6	67·5	76·6
United Kingdom	0·3	0·8	0·9	1·5	4·1	4·5	3·6	17·3	69·8	97·3	279·0	339·0
USA			0·2	0·5	1·1	1·7			10·8	33·5	69·4	117·8
	33·8	36·2	31·1	37·5	49·0	51·4	483·2	949·6	1,942·0	2,255·0	3,387·1	3,871·8
Finland	8·6	10·0	9·2	8·8	9·6	10·0	162·0	221·8	614·8	542·5	638·4	715·6
	42·4	46·2	40·3	46·3	58·6	61·4	645·2	1,171·4	2,556·8	2,797·5	4,025·5	4,587·4

Bulgaria	7·9	9·3	10·9	11·6	11·9	12·8	13·4	118·6	135·7	164·5	395·6	445·2	587·4
Cuba	7·0	7·4	7·6	8·1	8·8	9·1	9·4	92·2	114·2	175·8	135·8	288·2	375·2
Czechoslovakia	12·9	14·3	14·8	16·0	17·2	17·0	18·0	210·5	235·1	242·1	492·5	587·3	741·4
Germany	11·5	13·0	14·4	15·0	16·8	17·5	18·4	161·8	184·8	270·7	421·4	537·6	698·6
Hungary	5·5	6·3	6·7	7·6	8·4	8·7	9·3	93·8	112·9	140·8	308·8	377·3	502·8
Korea	0·4	0·6	0·9	1·1	1·1	0·9	1·1	13·8	16·8	24·7	26·6	43·7	47·3
Mongolia	0·3	0·3	0·3	0·4	0·4	0·4	0·5	10·9	12·1	12·5	13·1	25·7	35·0
Poland	11·1	12·3	11·9	13·3	14·1	15·3	16·0	182·2	214·0	243·8	524·3	591·5	802·3
Vietnam	0·2	0·2	0·3	0·4	0·4	0·6	1·0	8·6	10·0	11·3	14·2	14·4	27·8
Yugoslavia	3·4	3·9	3·8	4·4	4·9	4·8	51·3	99·3	248·6	272·5	318·5	343·5	
3rd World and Misc	2·6	3·0	3·7	5·3	5·5	6·5	60·2	67·6	71·6	77·9	84·0	87·0	91·0
REPORTED TOTAL							107·0	118·3	116·2	130	149		

*Includes West Berlin.

**1977 tonnage figures based on author's estimates.

***1972 1 ruble = $1.21
1973 1 ruble = 1.34

1974 1 ruble = $1.34
1975 1 ruble = 1.32
1976 1 ruble = 1.34

1977 1 ruble = $1.37
1978 1 ruble = $1.50

Sources:
Vneshniaia Torgovlia SSSR 1972 (Moscow, 1973). Hereafter VT SSSR. Also VT SSSR 1973, 1974, 1975, 1976, and 1977.

payback at that rate is quite reasonable. Hungarian sources estimate that their investment will be paid back in six to eight years.

The Soviets also benefit from the arrangement. While they are committed to supplying Eastern Europe with a minimum of 15·5 billion cubic meters a year, the total capacity of the pipeline is 28 billion cubic meters a year. That permits them to draw some extra gas for their own needs from the Orenburg–Soiuz pipeline, which they did even before the final pipeline was finished. In addition, as we shall see, the Soviets have every intention of utilizing the remaining capacity of the pipeline for sales to Western Europe as well as some additional sales to Eastern Europe.

No other joint CMEA project has been quite so spectacular or has involved so many partners. Several other projects had to do with non-energy raw materials, such as the cellulose plant at Ust-Illimsk, the asbestos plant at Kiembayev, and the steel plant at Kursk. Of more interest here, however, are the energy projects. These involve the 140 million ruble 750 kilovolt high-tension power-line from Vinnitsa in the USSR to Albertersa in Hungary.[27] The electricity to be transmitted will come in part from two nuclear energy plants built in the Soviet Union with East European help. Two other nuclear energy plants with a capacity of 4 million kilowatts are to be built in the Soviet Union with one-half of the output to be sent to Poland, Hungary, and Czechoslovakia via another 750 kilovolt line from Khmelnitskii in the Soviet Union to Zheshuv in Poland.[28] The Soviets have also enlisted East European help in building two assembly plants to manufacture atomic energy facilities.[29] One is to be built in the USSR and one in Czechoslovakia.[30]

East Europeans have also participated on a bilateral basis with the Soviet Union in the production of Soviet petroleum. Thus on 23 June 1966 the Czechs provided a 500 million ruble loan to the Soviet Union to facilitate oil exploration. In exchange the Czechs were guaranteed delivery of 5 million tons of petroleum a year from 1975 to 1984 over the level of 1970 deliveries.[31] Whether or not this loan was the sole cause for the increase that took place is hard to tell. But whereas Soviet exports of oil and petroleum products to Czechoslovakia in 1970 amounted to 10·5 million tons, by 1975 they had increased to 16 million tons and continued to increase in subsequent years.[32] (See Table 4.2.) Other countries involved in similar arrangements include East Germany, which agreed on 4 April 1967 to build a refinery for the Soviet Union, as well as Bulgaria and Hungary.[33]

Notice, by the way, that most of these agreements to help the USSR were signed in the mid-1960s, just after the Soviet Union began complaining about how it was being exploited by the East Europeans. The pressure to help the Soviet Union continues, however. In 1974,

Poland agreed to lay a 900 kilometer pipeline and build a pumping station in exchange for which the Soviets agreed to increase the deliveries of crude oil from the 7 million tons sent in 1972 to 11 million tons by 1975.[34] (See Table 4.2.) Overall the East Europeans had to put up 1·2 billion rubles to increase the flow of oil and gas from the Soviet Union from 1960 to 1974.[35] To that, of course, should be added the 2·2 to 2·4 billion ruble Orenburg pipeline for a total of close to $5 billion.[36]

RECAPTURING WESTERN MARKETS

Complicated as Soviet trade is with Eastern Europe, the Soviet trade relationship with Western Europe and the advanced capitalist world is even more complicated. Petroleum exports, with two exceptions, are on a straight buy-and-sell basis with no pressure on the importer to help the Soviet Union with an investment program. Those exceptions are in Sakhalin, where a Japanese–American group is involved in a joint venture with the Soviets to find oil, and in Siberia, where a Canadian company is exploring for petroleum. But except for these two instances, the sale of Soviet petroleum to the hard-currency world differs from the arrangements made with Eastern Europe because the Soviets do not necessarily expect Western companies to share or underwrite the investment expense of a project. They differ also in that in the years prior to 1973 the Soviets did not always find it easy to sell petroleum on capitalist markets.

In the Cold War era of the 1950s with its cheap and readily available supply of petroleum, Soviet oil sales outside the CMEA bloc were viewed as disruptive. The Soviets were regarded as trouble-makers, as indeed was often the case. In the eyes of many, their attempts to sell oil were treated as an economic and political plot. Non-communist buyers were warned that the Soviets were trying to bankrupt the Western companies at which time they would then jack up the prices and exercise political blackmail on those who had become dependent on Soviet oil and gas. While such charges were, on the whole, unfair, there is no doubt that the Soviets were interested in more than maximizing their profits.

With much of their capacity destroyed immediately after the war, the Soviets were initially not particularly motivated by thoughts of disrupting energy supplies in the West. Their first concern was reconstruction and satisfying their domestic needs. Yet, earlier than one would have expected, the Soviets began to re-establish their links with former capitalist customers. Of course, during the Cold War, trade volume never reached a very high level, but as early as 1946 the Soviets made a token delivery of 18,000 tons of petroleum to Finland and about 800 tons to Sweden. The following year Italy, Denmark, and the United Kingdom were added as customers. Each year new customers joined the list, even though it was not until 1954 that the Soviets had a net export balance in their petroleum trade.

Of course, the preponderant share of Soviet petroleum exports were directed to their communist allies, but even in the Cold War there were some substantial Soviet sales to traditional customers in the non-communist world. Exports to Finland totalled nearly 200,000 tons as early as 1949. That made Finland the Soviet Union's largest customer in both the communist and non-communist worlds. Except for China, Finland was also the largest importer in 1953 and 1956. The exports to Sweden, which began in 1946, continued at a moderate rate until 1953, after which they jumped to 269,000 tons and kept going. By 1959 they exceeded 1 million tons. In part this trade reflects the fact that Sweden was the only non-communist country to loan money to the Soviet Union after the Second World War.[37] Since it was not a member of NATO, Sweden regarded itself as a traditional neutral. Thus, not obligated by the NATO embargo, Sweden willingly bartered steel pipe to the Soviet Union in exchange for various goods, including petroleum. According to some reports, by the mid-1950s, the Soviet Union was supplying 50 per cent of Sweden's oil.[38]

The case of Italy is somewhat more intriguing. Italy, after all, had been one of the Soviet Union's best customers during the interwar years. Perhaps it was only natural, therefore, that as early as 1947 the Soviet Union resumed their exports. However, except for 1952 when Italy was second only to China as the largest importer of Soviet oil, export levels were not especially large. The significance of Soviet exports to Italy lay more in the fact that such trade existed at all with a NATO country.

Exports to Italy increased significantly after the closing of the Suez Canal in 1956, when imports of Soviet crude and refined products more than doubled and reached a total of half a million tons. In 1957 the maverick Italian oil company ENI (Ente Nazionale Idrocaburi – the National Hydrocarbon Agency) signed its first agreement with the Soviet Union. As a result by 1958 combined imports doubled again to 1 million tons.[39] Like the Rothschilds in the 1880s, Enrico Mattei, the head of ENI, was then engaged in a fight to secure his own supply sources of petroleum in order to free himself from dependence on the leading seven international oil companies. He personally stepped into the negotiations in October 1960 and again in 1963.[40] Already in 1959 imports of crude and refined products exceeded 3 million tons. That made Italy the largest Soviet customer, surpassing exports to any one communist country. When Mattei joined the negotiations, imports jumped again in 1960 to 4.7 million tons and rose every year until

1967, when they peaked at 11,979 thousand tons.[41] Even then the Soviet Union continued to export more crude and refined product to Italy than to anyone else until 1970 when combined Czech imports of refined product of 10·5 million tons finally exceeded those of the Italians.[42] At its peak in 1962–3, ENI was importing up to 38 per cent of its crude oil import needs from the Soviet Union.[43]

Such a radical reorientation in Italy's supply pattern was bound to be noticed and to distress its traditional suppliers. Nor were the seven sister oil companies or the producing countries unmoved by the fact that the Soviets offered Mattei this oil in a complicated barter contract that amounted to about $1·15 a barrel or one-third of the then prevailing Middle East price.[44] Moreover, the Italians were not the only ones who were being enticed by Soviet oil. The Soviets were beginning to open up major new markets in such traditional capitalist markets as Belgium, Greece, West Germany, Iceland, Israel, France, and Japan.

In the minds of some, this was all a part of a diabolical Soviet plot to destroy NATO. It was one thing to carve out Eastern Europe and China as one's market. It was another to release suddenly large quantities of petroleum outside the bloc. Nor did the Soviets restrict themselves to Western Europe. The Soviet Union also exported to the Third World, but there was worry enough just in Western Europe. By 1959 Soviet exports to the non-communist world were already 50 per cent higher than in the prewar peak years of 1932. As Table 2.3 indicates, overall Soviet exports in 1959 exceeded 20 million tons for the first time and in 1963 they exceeded 50 million tons (1 million barrels a day). By 1966, after the opening of the Druzba pipeline linking Soviet fields to Schwedt in East Germany and the Baltic seaports, exports reached about 75 million tons. Nor did the second closing of the Suez Canal in 1967 hurt Soviet sales opportunities. As Soviet exports continued to grow, the Soviets came to have a significant share of some markets. We already saw how important they had become in Italy. In addition, at one point or another in this period, Iceland received about three-quarters of its oil from the Soviet Union, Austria one-quarter, Greece about two-fifths, and Finland three-quarters.[45] By 1971, when the United Kingdom formally agreed to change its laws restricting Soviet petroleum imports, all the major European countries, the United States, and Japan were importing Soviet oil.

Inevitably exports of this magnitude began to have an impact on the world market. The reappearance of the Soviet Union as a major factor in the world petroleum market came at an awkward time. Raw material prices of commodities, including petroleum, in the immediate postwar era had fallen as new capacity began to increase faster than the increase in demand. The pattern was reversed when the Korean War brought a temporary surge in raw material prices that ended in the mid-1950s.

Having become accustomed to the higher prices, raw material producers, and the oil countries in particular, were distressed to see the demand for their products fall again. Even more upsetting was the sudden appearance of a new entrant in the petroleum market. To many of the established producers, the Soviet Union was the source of the price weakening. The drop in the posted price of petroleum in 1959 and again in 1960 was blamed by some directly on Soviet exports and price-cutting.[46]

Reaction was immediate. The oil companies began to organize a campaign to boycott Soviet exports. They became even more incensed when the Soviet Union managed to encroach on their markets in Cuba, India, Ceylon, and Guinea. Typically the Soviets would offer petroleum at a price considerably below the then prevailing level. That in itself was enough to embitter the Western companies, but to make the offer even more tempting, the Soviets were willing to accept payment in soft currency or bartered goods. Once assured of an alternative source of supply, these countries began to pressure the oil exporting countries for lower prices. When the oil exporters refused to match these prices, most of those countries chose to accept the Soviet offer. They soon discovered, however, that the storage and distribution network was completely in the hands of the Western companies which almost always refused to process anything but their own petroleum. Their equipment was not adapted to handle Soviet petroleum and its heavy sulphur content. Inevitably this resulted in an increase in the tension which in the case of Cuba and Ceylon led to the nationalization of the assets of the Western oil companies. Cuba went even further and began to import all its petroleum from the USSR.[47] After two or three similar episodes, the oil companies found that with a little effort they could, after all, process Soviet oil in the refineries, storage tanks, and gasoline stations.

Still, while Western companies found it expedient to compromise when forced, they were not to happy with what they regarded as the growing Soviet threat to their markets. Thus to prevent the recurrence of such episodes, the oil companies appealed to Congress and NATO to support them in their efforts to shut off Soviet exports. Western oil executives argued that the Soviet oil offensive was designed 'as a weapon with which they hoped to destroy the private oil industry', and by extension, Western democracy.[48] As the oil companies and others saw it, 'The Soviet Union is using every means to encourage state control of oil in free world countries and to incite the leaders of the developing nations against private oil industry. The facilities have already been expropriated in Ceylon and Cuba and the industry is under heavy pressure in many countries as a result of Soviet offers of oil aid.' Nor were the United States companies merely forced out. 'Once they [the Soviets] get the capability they can use oil for political objectives and always hold over the head of those getting it, the threat of shutting off their oil supplies which would cause disruption in their economies.'[49] Those Western buyers who persisted in their efforts to buy Soviet petroleum were openly accused of trying to sabotage NATO and the West.[50] What was needed was a co-ordinated drive to stop the Soviet economic offensive. This led NATO to propose ultimately that the Common Market countries hold their purchases of Soviet oil to 10 per cent of their total needs.[51]

Of even more far-reaching consequence was the reaction of the oil producing countries. Never entirely happy with the revenue and taxes they were collecting, their distress increased when the oil companies lowered the posted price in 1959 and 1960.[52] After considerable discussion and consultation, Saudi Arabia, Kuwait, Iran, Iraq, and Venezuela agreed that individual action alone would not suffice, and that a co-ordinated response was necessary. Thus in August 1960 they decided to organize themselves into what they called the Organization of Petroleum Exporting Countries, to see what they could do to increase oil prices.[53] In a very real sense, Soviet action helped to give birth to OPEC.

The Soviet reaction to all these developments was part delight and part bewilderment. On the one hand, the Soviets certainly recognized, even belatedly, that prices fell after the Korean War, which caused economic difficulties for the exporters in the less-developed countries.[54] They also acknowledged that a new seller in the market tends to destabilize prices; at least, that is what they now complain China is doing as it seeks to sell its petroleum.[55] Nevertheless, at the time the Soviets were somewhat less than distressed when their cut-rate oil served as a wedge to restrict or even push out the oil companies from Cuba, Ceylon, India, Guinea, and Ghana.[56] On the other hand, the Soviets could not help but be somewhat surprised by the Western reaction. All they were doing was competing in the conventional way by lowering prices. In addition, as we just pointed out, the Russians had been there first. Reminding their critics that except for the war and reconstruction they had a long history of oil exports, the Soviets asserted that all they were trying to do was reclaim their rightful market share. Although he was being slightly ingenious, E. P. Gurov, the head of Soiuznefeeksport, the Soviet Oil Exporting and Foreign Trade Organization, could justifiably claim at the meeting of the 2nd Arab Petroleum Congress in Beirut in October 1960, that 'We have not entered the market to upset the traditional position of other sellers. We just want to regain our legitimate position among the exporting nations – the position we held during the prewar period and which we were compelled to abandon, owing to the war, and to the need to rebuild our national industry in the immediate postwar period.'[57] Gurov also sought to show how the Soviet Union after all had supplied 48 per cent of Italy's crude oil and petroleum product needs in 1935. Hence, what was so threatening if the USSR supplied 18 per cent in 1962?[58]

Reclaiming their market share was not the sole reason for Soviet agressiveness in world energy markets, but it is true that many had forgotten how important Russia had been as an oil-exporting country both prior to the revolution and in the 1930s.

THE SOVIET MULTINATIONAL CORPORATION

Obviously with exports exceeding 100 million tons (2 MBD) by 1971, the Soviets had to have a relatively sophisticated organizational framework to handle a volume of this magnitude. Building on their experience prior to the Second World War, Soiuznefteexport, the foreign trade organization for petroleum exports, began to resurrect its overseas network of trading companies. (Appendix 1 reprints the charter for Soiuznefteexport.) One of its first moves was to take over two Finnish oil companies. Suomen Petrooli-Finska Petroleum was founded on 18 October 1932, but Soiuznefteexport did not become the main shareholder until 1947.[59] The following year Soiuznefteexport also assumed majority control over Trustivapaa Benziini, which had originally been founded on 4 September 1934. The name of the latter was changed to the commercially more attractive Teboil in 1966.[60]

Ownership of these two companies helps to explain why Finland began to import so much Soviet oil so early. Over the years, the operations of both companies has continued to grow. Suomen Petrooli imports and exports petroleum and distributes it in Eastern Finland where it has seventy filling stations, and owns two oil tanks and six oil depots with a capacity of 90,000 meters. Teboil has about three hundred filling stations and fifteen oil depots with a capacity of 515,000 cubic meters.

As in all such operations, ultimate control resides in Moscow with the responsible foreign trade organizations, in this case Soiuznefteexport. That is not to say the relationship is necessarily that straightforward and simple. Not that it matters, but officially shares are held in Suomen Petrooli, not only by Soiuznefteexport, but by Teboil, and another Finnish company called Konela. Teboil in turn is also owned by Soiuznefteexport, Suomen Petrooli, and Konela. Just to interlock the directorates, Konela is also owned by Avtoeksport (auto exports) and Traktorexport (tractor exports), which are Soviet foreign trade organizations, and, as one might have guessed, Suomen Petrooli and Teboil![61]

The second area of operations is the United Kingdom. Most of the marketing arm of the prewar Russian Oil Products Company (ROP) had been sold during the Second World War to Regents Oil, which in turn was purchased by Texaco Oil Company. As Soviet export capacity began to grow, Soiuznefteexport decided to re-enter the market, and in 1959 formed Nafta GB Ltd. Besides Soiuznefteexport, technically its shares are also owned by the Russian Oil Products Company, the English–Soviet Shipping Company, and Arkos, all enterprises controlled by Soviet foreign trade organizations.[62] While it has its own network of filling stations, which at one time totalled over 400, and storage facilities, Nafta also operates as a trader.[63] But some Nafta practices are highly creative. For example, when oil dealers in the United Kingdom were prohibited from importing Soviet crude oil, Nafta GB found itself importing crude oil from that great oil producing state, Finland![64] For those who were under the impression that the Finns had no crude of their own worth mentioning, the puzzle becomes a little clearer when it is remembered that Finland receives over three-quarters of its oil from the Soviet Union.

While it was formed after Nafta GB, Nafta B (Belgium) has become the most ambitious and sophisticated of Soiuznefteexport's affiliates. Its Soviet staff is Westernized and its Belgian employees are highly professional. For example, the Belgian public relations director was hired away from Gulf Oil. Nafta B is also one of the few Soviet operations in Western Europe that is owned in part (albeit a steadily diminishing part) by local non-Soviet investors. Founded on 1 December 1967, Nafta B is owned by the Soviet foreign trade organizations Soiuznefteexport, Soiuzpromexport (industrial exports), Soiuzchemexport (chemical exports), Avtoexport, and three Belgian companies, which are not, at least not openly, affiliates of Soviet foreign trade organizations. They are Belgian Bunkering and Stevedoring, Antuane Vloberg, and De Murchie.[65] Initially the capital stock of Nafta amounted to only 3 million Belgian francs.[66] Subsequently, this was increased to 133 million Belgian francs, and in the process the share of the Belgian parties was reduced from about 40 per cent to 1 per cent.

Easily accessible to the Soviet pipeline terminus on the Baltic Sea at Ventspils, Latvia, Nafta B is the main link in the Soviet petroleum trading process. This is no minor matter, since approximately 30 per cent of all Soviet petroleum exports or close to $2 billion worth of sales are sold to the non-communist world through Soviet joint stock companies such as Nafta B.[67]

Not surprisingly, therefore, Nafta B has a very modern and sophisticated oil storage facility with a capacity of 850,000 cubic meters in the port of Antwerp. Reportedly, this facility cost 750 million French francs and was financed by Belgian banks. Subsequently, in 1973, Nafta added to this capacity by constructing additional storage facilities

Table 4.3 Consumption of Petroleum in the USSR (Production, Export, and Import) (million metric tons)

		Export			Import			Net	
	Production	Crude	Refined Product	Combined	Crude	Refined Product	Combined	Export	Consumption
1929	13·7	0·3	3·5	3·9	—	—	—	3·9	9·8
1930	18·5	0·3	4·4	4·7	—	—	—	4·7	13·8
1931	22·4	0·4	4·8	5·2	—	—	—	5·2	17·2
1932	21·4	0·5	5·6	6·1	—	—	—	6·1	15·3
1933	22·5	0·5	4·4	4·9	—	—	—	4·9	17·6
1934	24·2	0·5	3·9	4·4	—	—	—	4·3	19·9
1935	25·2	0·2	3·2	3·4	—	—	—	3·4	21·8
1936	27·4	0·2	2·5	2·7	—	—	—	2·7	24·7
1937	28·5	0·1	1·9	1·9	—	0·1	0·1	1·8	26·7
1938	30·2	0·2	1·2	1·4	—	0·1	0·1	1·3	28·9
1939	30·3	—	0·5	0·5	—	0·1	0·1	0·4	29·9
1940	31·1	—	0·9	0·9	—	0·1	0·1	0·8	30·3
1941									
1942									
1943									
1944									
1945	19·4	—	0·5	0·5	—	0·9	0·9		
1946	21·7	0·1	0·8	0·8	0·1	0·5	0·6	(0·4)	22·1
1947	26·0	0·1	0·7	0·7	0·1	0·8	0·9	0·2	25·8
1948	29·2	0·1	0·8	0·9	0·1			(0·2)	29·4
1949	33·4	0·1	0·8	0·9	0·1	1·7	1·8	(0·9)	34·3
1950	37·9	0·3	0·8	1·1	0·3	2·3	2·6	(1·5)	39·4
1951	42·3	0·9	1·6	2·5	0·1	2·6	2·7	(0·2)	42·5

Year									
1952	47.3	1.3	1.8	3.1	0.2	3.6	3.8	(0.7)	48.0
1953	52.8	1.5	2.7	4.2	0.1	4.6	4.7	(0.5)	53.3
1954	59.3	2.1	4.4	6.5	0.2	3.8	4.0	2.2	57.1
1955	70.8	2.9	5.1	8.0	0.6	3.9	4.5	3.5	67.3
1956	83.3	3.9	6.2	10.1	1.5	3.8	5.3	4.8	79.0
1957	98.3	5.9	7.8	13.7	1.3	2.9	4.2	9.5	88.8
1958	113.2	9.1	9.0	18.1	1.1	3.2	4.3	13.8	99.4
1959	130.0	12.5	12.9	25.4	1.1	3.3	4.4	21.0	109.0
1960	147.9	17.8	15.4	33.2	1.2	3.2	4.4	28.8	119.1
1961	166.1	23.4	17.8	41.2	0.9	2.7	3.6	37.6	128.5
1962	186.2	26.3	19.1	45.4	0.5	2.3	2.8	42.6	143.6
1963	206.1	30.2	21.1	51.3	0.5	2.3	2.8	48.5	157.6
1964	223.6	36.7	19.9	56.6	—	2.1	2.1	54.5	169.1
1965	242.9	43.4	21.0	64.4	—	1.9	1.9	62.5	180.4
1966	265.1	50.3	23.3	73.3	—	1.7	1.7	71.6	193.5
1967	288.1	54.1	24.9	79.0	0.1	1.4	1.5	77.5	210.6
1968	309.2	59.2	27.0	86.2	0.9	1.1	2.0	84.2	225.0
1969	328.4	63.9	26.9	90.8	2.5	1.1	3.6	87.2	241.2
1970	353.0	66.8	29.0	95.8	3.5	1.1	4.6	91.2	261.8
1971	377.1	74.8	30.3	105.1	5.1	1.5	6.6	98.5	278.6
1972	400.4	76.2	30.8	107.0	7.8	1.3	9.1	97.9	302.5
1973	429.0	85.3	33.0	118.3	13.2	1.5	14.7	103.6	325.4
1974	458.9	80.6	35.6	116.2	4.4	1.0	5.4	110.8	348.1
1975	491.0	93.1	37.3	130.4	6.5	1.1	7.6	122.8	368.2
1976	520.0	110.8	37.7	148.5	6.4	0.8	7.2	141.3	378.7
1977	546.0	155–159							
1978	572.0								
1979	586.0								

in Liege and Brussels with a capacity of 600,000 tons. Technically, in Nafta B, the Soviet Union has created a fully integrated multinational firm. It has its own sources of supply, its own tankers and pipelines, its own tank farms, its own wholesale operation in at least three other foreign countries, and its own filling stations. The network continues to grow. Branch offices were opened in Denmark in 1975, in Italy in 1976, and in Germany shortly thereafter.[68] However, not all the efforts have been successful. The Soviets have tried unsuccessfully to buy or build a refinery, first in Belgium near their Antwerp storage facility, then in the United States on the Gulf Coast, and near Le Havre. In fact, the contract was actually signed for the construction of the French project, but in each instance construction was ultimately cancelled.[69]

SOVIET PETROLEUM IMPORTS

Though Soiuznefteexport's main function is the export of petroleum, as we saw, the Soviet Union also imports some petroleum and was an importer in the immediate postwar era. As Tables 4.3 and 4.4 indicate, until the 1970s refined products were the major import item. Initially Soviet refinery capacity and sophistication left something to be desired. Accordingly the Soviets supplemented their own efforts with imports which until 1965 amounted to a minimum of 2 million tons a year. Rumania was the main supplier, but the Soviet Union imported small amounts from a variety of suppliers, including some American companies. (See Table 4.4.) As Soviet productive capabilities have improved, however, the Soviets have gradually reduced the amount of refined products they have had to import, so that imports now total less than 1 million tons a year.

The pattern of crude oil imports is somewhat different. There were minute amounts imported after the Second World War until the mid-1950s. Then in 1955 the Soviets started importing large quantities of crude oil from, of all places, Austria.[70] They took over an oil field concession in Austria as part of their booty from the Second World War. They returned the field to Austria in 1955 but forced the Austrians to compensate them in kind with larger petroleum deliveries in exchange for giving them back their own oil fields. For the next five years the Soviets imported over 1 million tons a year, almost all from Austria. These imports dropped gradually to about half a million tons, and then ceased completely in 1964 when the Austrian compensation payments had been completed.

When the Soviet Union began to direct its foreign aid and arms sales to developing countries like Algeria, Egypt, and Syria, imports of crude resumed, although almost entirely as aid and weapons repayment. To the extent that this petroleum was re-exported by the Soviets or freed

Table 4.4   *Crude Oil and Oil Product Imports of the USSR (thousand tons)*

	1972	1973	1974	1975	1976	1977**
Crude						
Algeria	570	NOR*	NOR	984	NOR	
Iraq	4,084	11,010	3,888	5,304	5,821	4,650
Libya	1,867	1,713	NOR	NOR	NOR	1,000
Egypt	971	209	172	211	154	
Syria	315	247	330	NOR	450	
People's Democracy of Yemen	NOR	NOR	0·3	0·1	0·1	
Subtotal	7,807	13,179	4,093	6,499	6,426	5,600
Reported crude total	7,800	13,200	4,390	6,499	6,425	
Refined product						
Rumania	352	522	455	492	155	
United States	NOR	NOR	2	42	49	
Poland	495	417	84	NOR	NOR	
Subtotal	847	939	541	534	204	
Refined product total	1,300	1,500	996	1,060	797	
Combined total	9,100	14,700	5,386	7,559	7,222	

* NOR = None Officially Reported.
** 1977 figures based on author's estimates.
Source:
Various issues of *VT SSSR*.

Soviet oil for export, such arrangements made good sense, particularly when the Soviet sale was made in hard currency. Some sales were also arranged as swap-deals, particularly after the 1967 Six Day War and the closing of the Suez Canal. For example, among the deals reportedly made was one whereby British Petroleum sold 700,000 kiloliters of Abu Dhabi oil to Soiuznefteexport, so the Russians could fulfill their delivery commitments to Idemitsu Kosan in Japan and to some other companies in Ceylon without having to go all around South Africa.[71] In exchange the British took delivery on Baku oil which they then were able to ship to their European customers, thereby also avoiding the South African detour. Swiss and Swedish companies have also reported being engaged in similar swaps.[72] The reported import into the Soviet Union of refined oil products from Kuwait in 1971 may reflect yet another such transaction.[73]

The third closing of the Canal during the Yom Kippur War of 1973 seems to have provided a stimulus for more swap-deals, this time with British Petroleum (BP), a French company, and Japanese companies.[74] The volume of Soviet imports from the Middle East increased and rose to 13 million tons in 1973 (mostly from Iraq), then fell back, and seemed to level off at 5 to 6 million tons a year. (See Table 4.4.) Iraq became the main supplier and apparently the main swapper. The Soviets have even come to acknowledge that some of their sales are being made with Iraqi oil.[75] Finally, in an arrangement that has nothing to do with the Suez Canal, only good sense and the elimination of cross hauling, the Soviets have sought to arrange a swap with both Venezuela and Mexico. They will supply Cuba with their oil, while the Soviet Union fulfills Mexico's and Venezuela's commitments in Europe.[76]

Oh, how times have changed! While once both the oil companies and the oil producers viewed Soviet petroleum as an economic threat to their very existence, by the late 1960s they had begun to work cordially on co-ordinated ventures as well as to buy and sell to each other.

NATURAL GAS EXPORTS

Except for a very brief note in the preceding chapter, little has been said about gas exports, because there were none of any consequence until the 1940s. It took until then to find a practical way to put the gas to use and to transport it. Eventually a few pipelines were built from the Dashava gas fields in the western Ukraine, heading both east to Moscow and west to Lvov.[77] Once the war was over, the Soviets started to ship small quantities of gas to Poland from the Ukrainian fields as early as 1946.[78] This was not too surprising because the gas fields were located in what originally had been Poland but only became the western Ukraine when the Soviets forced the Poles to give up the land during the Nazi–Soviet Pact years prior to the Nazi invasion. Thus the pipeline from the field to Poland was already in place. What had been an internal transaction became an international export.

Until 1958 these shipments never amounted to more than 200 million cubic meters. The really big change came in 1967 with the opening of the Bratstvo (Brotherhood) pipeline to Bratislava, Czechoslovakia. This pipeline made it possible to increase exports to Poland to 1 billion cubic meters and to begin exporting to Czechoslovakia. On 1 September 1978 the following year, Austria was linked up to the pipeline. By the mid-1970s pipeline capacity had been increased and extended, so that Bulgaria, the German Democratic Republic, West Germany, Hungary, France, Italy, and Finland had all become regular and substantial consumers. (See Table 4.1.) By 1977 West Germany and Italy, for example, were importing close to 6 billion cubic meters of Soviet gas a year.

The West European sales almost always have been predicated on a massive pipe for gas barter arrangement, whereby the Soviets have committed themselves to repay the deliveries of pipe with gas.[79] We saw earlier how much pipe the Soviets have had to import. The bulk of it was utilized on those pipelines. But the gas exported has eventually turned out to be equally important for those who consume it. By 1980, Soviet gas constituted about 20 per cent of the gas imported by the West European customers.

Having seen the potential, the Soviets eagerly began to consider other ways of exploiting their ever-growing discoveries of natural gas. But because of the remoteness of most of their deposits, this was easier dreamed than done. One approach has been to encourage the involvement of American companies, primarily in the Urengoi fields in west Siberia, the middle Viliui fields farther east near Yakutsk and the fields on Sakhalin, north of Japan. Called the North Star Project, the gas from the Urengoi field was to be transported in a pipeline to Murmansk about 1,600 miles away on the Barents Sea. It would be liquified there and then shipped in cryogenic tankers through the Atlantic to the United States. In exchange for building the pipeline and liquification plant, the American consortium made up of Tenneco, Texas Eastern Gas, and Brown & Root was to be supplied with 2 billion cubic meters a year for twenty-five years.

The Yakutia project based on the Viliui fields was assigned in June 1973 to El Paso Company with a 75 per cent interest and Occidental Petroleum with 25 per cent. The Bechtel Corporation was also a participant. They, in turn, have brought in some Japanese companies to take over a 50 per cent interest in the whole project. In this instance, a pipeline 2,000 miles long would be built east to Nakhodka, where, as at Murmansk, 10 to 20 billion cubic meters of gas a year for twenty-five years would be liquified and sent in refrigerated tankers, this time to Japan and the west coast of the United States. Letters of intent have been signed, but further work is contingent on the Soviets being able to demonstrate that the Yakutia field contains 1 trillion cubic meters of commercially exploitable reserves, and on the American side obtaining permission to import liquified gas and the money to underwrite this multi-billion dollar project.[80] As of 1979, proved reserves were estimated to be over 800 billion cubic meters, still about 20 per cent short of the necessary minimum. The expectation was that the remaining reserves would be proven in one or two years.

The prospects for financing were much less optimistic. The North Star and Yakutia projects would cost upwards of $6 billion apiece in 1979 dollars. The only way to raise such sums of money in the United States would be to involve the Export-Import Bank. However, the Church Amendment to the Stevenson Amendment to the

Export-Import Bank Act of 1974 and the Jackson–Vanik Amendment
to the 1974 Trade Act have restricted American credits for research
and exploration of energy projects to $40 million. Furthermore, since
none of the $40 million may be used for production, processing, or
distribution of Soviet fossil fuel energy, Export-Import Bank credits
can hardly be counted on to cover much of the $10 billion credit
requirement.[81]

Despite the staggering costs involved and the restrictions on
financing, neither the North Star nor the Yakutia project is dead yet. In
1975 the American companies went to Western Europe to seek
financing. They found government and private investors in France,
West Germany, and the United Kingdom who expressed willingness to
put up the money for the North Star project. It was thereupon
redesignated North Star International. Under the new scheme there
would be no need for United States credits.[82] The Yakutia project is also
alive, with the Japanese reportedly interested in sharing the costs.

The third project, on the Island of Sakhalin, is much less spectacular.
This harkens back to pre-Second World War days when the Japanese
had a concession to develop liquid energy there. The Japanese continue
as the major partner, but Gulf Oil has at least a 7 per cent share in the
project, which involves a search for both oil and gas.[83] As in the other
gas projects, the capital requirements are enormous and, as at Yakutia,
there is some uncertainty as to just how extensive the gas reserves are.
Nor were the Japanese and Gulf Oil reassured when Premier Kosygin
warned the Japanese that instead of the 60 billion cubic meters the
Japanese had been led to believe they would find, there may be only 16
billion cubic meters in reserve.[84] None the less, the Japanese as of
mid-1979 had put up $150 million in loans to finance their exploratory
work.[85] They were, no doubt encouraged in their efforts by the fact that
oil was found in late 1977, furthering expectations that gas would also
be discovered.[86] The North Star International, the Yakutia, or the
Sakhalin ventures are not necessarily the most complicated ever
undertaken by the Soviets. Just as in the case of Soiuznefteexport, the
Western observer is impressed by the sophistication of Soviet energy
transactions. The counterpart to the Oil Foreign Trade Organization in
the gas business is Soiuzgazexport.[87] (Its charter is reprinted in
Appendix 2.) An example of its flexibility and ingenuity can be seen in
the way it has arranged to sell gas to France. Since there is as yet no
pipeline tying in France to the Soviet network, this turns out to be no
mean accomplishment. Instead of waiting for the pipeline to be built,
the Russians designed a four-way swap deal. Thus the Soviet gas sold to
France actually goes to Italy instead, which is already tied into the
Soviet pipeline system. In exchange, the Italians have asked the Dutch
to deliver the Dutch natural gas the Italians had ordered to France.

A transaction that appeared to be equally ingenious, at least on paper, involved even more countries. In this instance the Soviet Union joined together in July 1977 with the Czechs, the West Germans, the Austrians, the French, and the Iranians. Under the contract the Soviets agreed that they would build a 600-mile natural gas pipeline from Iran to the Soviet Union in return for which the Soviet Union would be paid with 1 million tons of Iranian crude oil.[88] Beginning in January 1981 and reaching full capacity in 1984, this pipeline would carry 17·2 billion cubic meters of gas a year until 2001 to the Soviet Union. According to the plan, the Soviet Union would then be able to use 17·2 billion cubic meters for its own consumers in the Caucasus and Central Asia and the Donbass in the southern Ukraine.[89] It would, however, take 13·5 billion cubic meters of natural gas from other deposits in the Soviet Union and ship this gas to Czechoslovakia, Austria, West Germany, and France, thus keeping about 3·5 billion cubic meters a year.[90] Some of this would go to the Soviet Union free as a transit charge. The Czechs in turn would draw off some 2 billion cubic meters, part of which would be a transit fee, and send off 1·86 billion cubic meters to OMV in Austria, 5·7 billion cubic meters to Ruhrgas in Germany, and 3·8 billion cubic meters to Gaz de France.[91] Apparently Iranian exports were originally set at 13·6 billion cubic meters, but were increased to allow an extra 2 billion cubic meters for the Czechs.

Note the beauty of the scheme! The Soviets would receive 1 million tons of oil for building the Iranian pipeline, 3·5 billion cubic meters of gas in part as a transit fee, and the entire cost of the project would be underwritten by the West Europeans. Nor would the Soviets have to build a very long pipeline. At most they would have to build a 760-mile line from the Iranian border to near Rostov in the Donbass. They already have surplus pipeline capacity to the West. With the completion of the Orenburg–Soiuz pipeline and its capacity of 28 billion cubic meters, the Soviets could theoretically satisfy their commitment to the East Europeans for 15·5 billion cubic meters and still fulfill virtually all of their obligations under the trilateral deal with Iran and Western Europe. At least, that was the plan.

Nor did the plan seem unrealistic. After all, the Soviets had been importing Iranian gas since 1970. Under an agreement signed in 1966, the Iranians and Soviets agreed to build a pipeline from the southern oil fields in Iran to the Soviet border at Azerbaidzhan.[92] Called the Iranian gas trunkline (IGAT I) (the new line involving Austria, Germany and France was to be called IGAT II), it opened on schedule in October 1970, and by 1972 Iranian shipments to the Soviet Union began to average 8 to 9 billion cubic meters of natural gas a year. This turned out to be a mutually advantageous arrangement. The Soviets took the gas as repayment for the steel mill which they had built for Iran at Isfahan and used the imported gas to increase gas supplies in Azerbaidzhan, Georgia, and Armenia, which hitherto had had to rely entirely on the inadequate quantity being produced at Baku. In turn the Iranians were able to derive some benefit from the by-product gas released in their oil extraction operations. Previously it was simply flared off as waste. Instead, beginning in 1970, it served as a means of repaying a Soviet debt and a source of energy for the Iranians located along the pipeline who drew off another 3 billion cubic meters of gas for their own use.[93]

Although not in as complicated a fashion, the Soviets also imported gas from Afghanistan. Like the Iranians, the Afghans used this gas as a form of repayment to the Soviet Union for foreign aid. Shipments from Afghanistan began even earlier, in 1967, and by 1970 averaged about 2·5 billion cubic meters a year.

For a time, from 1970 to 1973, the combined Afghan-Iranian imports exceeded Soviet gas exports. That led some observers to conclude incorrectly that the Soviets had encountered energy problems and had become dependent on natural gas imports.[94] Actually, the Soviets were awaiting the opening of the Bratstvo pipeline after which exports quickly exceeded imports.

Not to be outdone by their counterparts in Soiuznefteexport, the managers of Soiuzgazexport have also adroitly manipulated their prices. When it appears that their customers have no easily accessible alternative sources of supply, the Soviets generally price their natural gas at relatively high prices. As Table 4.5 shows, on the whole the East European countries, as well as Austria and Finland, which generally do

Table 4.5 *Prices of Soviet Natural Gas*
(kopecks per cubic meter)

	1972	1973	1974	1975	1976
Austria	11·6	10·5	13·6	30·3	33·4
Bulgaria			15·2	29·4	33·3
East Germany			15·1	15·1	27·7
Hungary				29·8	33·9
Italy			8·0	16·4	14·0
Poland	13·8	13·8	13·9	28·0	32·0
West Germany		14·6	14·1	17·8	22·7
Finland			47·7	46·8	47·3
France					25·6
Czechoslovakia	14·7	14·7	15·3	25·5	34·5
Imports from					
Afghanistan	5·0	5·4	6·8	12·4	12·6
Iran	6·2	7·8	14·6	15·3	15·6

not have the option of buying gas from the Netherlands or Algeria, have paid the highest prices. The pattern is somewhat mixed prior to 1976, but in 1977 all those countries (except for East Germany) paid 32 kopeks or more per cubic meter. All the others who had alternatives were charged only 26 kopeks or less. When it can, the Soviet Union will pay little but charge much. Thus it was able to export gas at a price double what it paid to the Afghans and Iranians for their imports. Since both the export and import price according to Soviet practice was f.o.b. the Soviet border, the price paid by the West Europeans was even higher. It looks as if Soiuzgazexport has negotiated well.

While it is very misleading to suggest that the Soviets manage to come out ahead in every negotiation, it is none the less striking just how well they do. After all, the international market for petroleum is not for amateurs nor for those unaccustomed to the market system, particularly its faster moving operations. How is it then that those born and bred in the turgid, overprotected, and restricted world of the planner and bureaucrat, which produces the irrational responses we considered earlier, do so well in the heady, everchanging atmosphere of the speculator and arbitrater. When asked that very question, one Soviet specialist replied only half in jest: 'It's very simple. Soviet managers are carefully trained to ensure an adequate flow of supplies and never to deviate from the plan. These are just the opposite talents needed in the capitalist world for Soiuznefteexport and Soiuzgazexport. So when we send our Soviet-trained managers out into the capitalist world, we simply tell them to do everything just the opposite that they would do at home!'

CONCLUSION

For Soviet oil and gas traders, the years from the Second World War to the 1973 Middle East War were marked by dramatic changes. After initially being regarded as an interloper and helping to spark the creation of OPEC, the Soviet Union gradually found itself winning acceptance in the world petroleum market. While this acceptance is due more than anything else to the gradual tightening of the world market for petroleum, it also shows that, given the opportunity, the Soviets will generally behave like their capitalist counterparts. We saw that when market conditions allowed, they paid low prices for imports and formed multinational affiliates which were not adverse to monopolizing the market and charging high prices.

Yet the Soviets derive no special pleasure from cutting prices. Normally they cut prices only when necessary to carve out a share of the market for themselves. This is well illustrated by the story told by the late chairman of Eastern Gas and Fuel Associates, Eli Goldston, whose company along with the Russians sold coal regularly to the Japanese. At the negotiating sessions, he noted that no matter what price he bid, the Russians would always follow up his price with a lower bid of about 50 cents a ton. One evening before the bidding he met his Russian counterpart at a bar and challenged him to enter the first bid the next day and let Goldston go second. The Russian readily agreed, replying somewhat to Goldston's surprise that the Soviets had already calculated what they would bid the next day. 'What is it?' Goldston asked. 'Fifty cents below your price!'

CHAPTER 4: NOTES

1 Anthony C. Sutton, *Western Technology and Soviet Economic Development, Vol. III, 1945 to 1965* (Stanford: Hoover Institution Press, 1973), p. 39.
2 Marshall I. Goldman, *Soviet Foreign Aid* (New York: Praeger, 1967), pp. 14–19.
3 *VT SSSR*, 1918–66, pp. 148–9.
4 Marshall I. Goldman, *Détente and Dollars* (New York: Basic Books, 1975), p. 23.
5 *VT SSSR*, 1918–66, pp. 126–9.
6 *VT SSSR*, 1956, p. 116; Samuel Pisar, *Commerce and Co-Existence* (New York: McGraw-Hill, 1970), pp. 182–275.
7 *VT SSSR*, 1918–66, pp. 126–7.
8 Sovet Ekonomicheskoi Vzaimopomoshchi, *Statisticheskii ezhegodnik 1971* (Moscow: Statistika, 1971) (hereafter SEV and appropriate year), p. 79.
9 *VT SSSR*, 1976, p. 196; SEV, 1978, pp. 76 and 388.
10 Jeremy Russell, *Energy as a Factor in Soviet Foreign Policy* (Westmead, England, and Lexington, Mass.: Saxon House, Lexington Books, 1976), pp. 11, 14, 91.
11 CIA Gas, July 1978, p. 61.
12 Horst Mendershausen, 'Terms of trade between the Soviet Union and the smaller communist countries, 1955–57', *Review of Economics and Statistics*, May 1959; 'The terms of Soviet satellite trade: a broadened analysis', ibid., May 1960; 'Mutual price discrimination in Soviet bloc trade', ibid., November 1962.
13 Franklyn Holzman, 'Soviet foreign trade pricing and the question of discrimination', ibid., November 1962, and 'More on Soviet trade bloc discrimination', *Soviet Studies*, July 1965, p. 44.
14 *VT SSSR*, 1960, pp. 84, 100, 116.
15 E. Nukhovich, 'Ekonomicheskoe sotrudnichestvo SSSR s osvobodivshimisia stranami i burzhuaznye kritiki', *Voprosy ekonomiki*, October 1966, pp. 83, 85, 86; N. Volkov, 'Struktura vzaimnoi torgovli stran SEV', *Vneshniaia torgovlia*, December 1966, pp. 10, 12.
16 *Foreign Trade*, June 1973, p. 14; G. Prokhorov, 'Mirovaia sistema sotsializma i osvobodivshiesia strany', *Voprosy ekonomiki*, November 1965, pp. 84, 85; O. Bogomolov, 'Khoziaistvennye reformy i ekonomicheskoe sotrudnichestvo sotsialisticheskikh stran', *Voprosy ekonomiki*, February 1966, pp. 85, 86; M. Sladkovskii, 'XXII s'ezd KPSS i problemy ekonomicheskogo sotrudnichestva sotsialisticheskikh stran', *Voprosy ekonomiki*, April 1966, p. 96.
17 *Foreign Trade*, June 1973, p. 14.
18 *Pravda*, 29 July 1968, p. 7; *Foreign Trade*, January 1971, p. 6, July 1973, p. 34; A. I. Zubkov, 'SSSR i reshenie toplivno-energeticheskoi i syr'evoi

problemy v stranakh SEV', *Istoriia SSSR*, no. 1, 1976, pp. 54, 58; A. Lalaianitz, 'Dva obshchestva – dve linin v razvitii toplivo-energeticheskogo kompleksa', *Planovoe khoziaistvo*, April 1975, p. 29.

19 CIA Gas, July 1978, p. 53; *Review of Sino-Soviet Oil*, October 1978, p. 41; *New York Times*, 30 October 1976, p. 4.
20 *Review of Sino-Soviet Oil*, December 1977, p. 71.
21 *Review of Sino-Soviet Oil*, December 1977, p. 54.
22 *Ekonomicheskaia gazeta*, no. 35, August 1978, p. 2.
23 CIA Gas, July 1978, p. 53.
24 *Review of Sino-Soviet Oil*, October 1976, p. 55.
25 *Sotsialisticheskaia industriia*, 29 September 1978, p. 3; *Soviet News*, 23 January 1979, p. 3.
26 *Review of Sino-Soviet Oil*, November 1978, p. 55.
27 *Moscow Narodny Bank Bulletin*, 25 August 1976, p. 12.
28 *Ekonomicheskaia gazeta*, no. 24, June 1979, p. 20; *Foreign Trade*, November 1978, p. 6.
29 V. Novikov, 'K novomu pod'emu otchestvennogo mashinostroeniia', *Kommunist*, no. 3, February 1979, p. 33.
30 *New York Times*, 30 June 1979, p. 28.
31 *Foreign Trade*, March 1978, p. 5; Zubkov, p. 60.
32 *VT SSSR*, 1970, p. 198; *Current Digest of the Soviet Press*, 20 October 1976, p. 12; S. Pomazanov and A. Iakushin, 'Razvitie integratsionnykh protsessov v energetike stran SEV', *Voprosy ekonomiki*, June 1976, p. 77; Zubkov, p. 60.
33 *Moscow Narodny Bank Bulletin*, October 25, 1976, p. 12; Zubkov, p. 60; Pomazanov and Iakushin, p. 77.
34 *Foreign Trade*, November 1975, p. 6. *Izvestiia*, 12 December 1974, p. 4; Zubkov, p. 60; Pomazanov and Iakushin, p. 77.
35 Lalaiantz, p. 29.
36 Stern, p. 14.
37 A. G. Zverev, *Finansy i sotsialistichekoi stroitel'stvo* (Moscow: Gosfinizdat, 1957), p. 326.
38 Park, p. 45.
39 *VT SSSR*, 1918–1966, pp. 126–7; Arthur Jay Klinghoffer, *The Soviet Union and International Oil Politics* (New York: Columbia University Press, 1977), p. 221.
40 *Foreign Trade*, September 1967, p. 18.
41 *VT SSSR*, 1968, p. 156.
42 *VT SSSR*, 1970, p. 68. Dow Votaw notes the large volume of Italian imports but mistakenly assumes that this made the Italians the largest importer only outside 'the Iron Curtain'. Votaw, *The Six-Legged Dog, Mattei and ENI, A Study in Power* (Berkeley: University of California Press, 1964), p. 5.
43 ibid., p. 5; P. H. Frankel, *Mattei: Oil and Coal Power Politics* (New York: Praeger, 1966), p. 138. Frankel says it was only 16 per cent, but this may refer to crude and refined products. At the other extreme Klinghoffer says that in 1961 ENI received 72 per cent of its entire oil supply from the Soviet Union and that as a whole Italy received 23 per cent. Klinghoffer, p. 221. Since ENI was also importing from Iran at the time, Votaw's figures seem the more appropriate ones.
44 Votaw, p. 36.
45 Klinghoffer, pp. 64–5; Goldman, *Foreign Trade*, p. 89.
46 Hearings Before the Sub-Committee to Investigate the Administration of the Internal Security Act and Other Internal Security Laws of the Committee on the Judiciary, United States Senate, *Exports of Strategic Materials to the USSR and Other Soviet Bloc Countries*, Eighty-Seventh Congress, Second Session

(Washington, DC, US Government Printing Office, 1963), Pt III, p. 382 (hereafter referred to as *Senator Keating Report*).

47 *Foreign Trade*, September 1967, p. 18.
48 ibid., p. 406; National Petroleum Council, *Impact of Oil Exports from the Soviet Bloc*, Vol. I (Washington, DC: National Petroleum Council, 1962), p. 37.
49 *Senator Keating Report*, Pt III, pp. 380, 405.
50 Frankel, p. 40; Votaw, p. 22.
51 *Foreign Trade*, September 1976, p. 19.
52 M. A. Adelman, *The World Petroleum Market* (Baltimore: Johns Hopkins University Press, 1972), pp. 167, 207.
53 Klinghoffer, p. 68, Stobaugh and Yergin, p. 24.
54 R. N. Andreasian and A. D. Kaziukov, *OPEC Mire Nefti* (Moscow: Nauka, 1978), p. 148.
55 *Ekonomicheskaia gazeta*, no. 42, October 1978, p. 21.
56 Marshall I. Goldman, *Soviet Foreign Aid* (New York: Praeger, 1967), pp. 96–7.
57 Cited in Neil H. Jacoby, *Multi-National Oil* (New York: Macmillan, 1974), p. 165.
58 *Foreign Trade*, September 1967, p. 17.
59 *Foreign Trade*, October 1975, pp. 60–71.
60 Central Intelligence Agency, *Soviet Commercial Operations in the West*, ER 77-10486 (Washington, DC, September 1977), p. 13.
61 *Foreign Trade*, December 1975, p. 54.
62 ibid., April 1977, p. 59.
63 *Moscow Narodny Bank Bulletin*, 14 February 1971, p. 2; *Petroleum Press Service*, July 1971, p. 267.
64 *Moscow Narodny Bank Bulletin*, 4 February 1971, p. 2.
65 *Foreign Trade*, January 1976, pp. 60–1.
66 *Moscow Narodny Bank Bulletin*, 14 January 1971, p. 2; *Petroleum Press Service*, January 1971, p. 29.
67 *Foreign Trade*, May 1979, p. 6.
68 ibid., January 1976, p. 61, October 1978, p. 28; *New York Times*, 8 July 1979, financial section, p. 1.
69 *Moscow Narodny Bank Bulletin*, 28 February 1973, p. 4; *Soviet News*, 3 April 1973, p. 153; *New York Times*, 21 June 1971, p. 45.
70 Klinghoffer, p. 183; *VT SSSR*, 1918–1966, pp. 162–5.
71 *East-West Trade*, May 1968, p. 62, September 1968, p. 68; *Petroleum Press Service*, June 1968, p. 206.
72 *New York Times*, 15 May 1968, p. 61; *Petroleum Press Service*, May 1969, p. 165.
73 ibid., April 1970, p. 154, November 1970, p. 429; *VT SSSR*, 1971, p. 253; Klinghoffer, p. 101.
74 *Moscow Narodny Bank Bulletin*, 29 January 1975, p. 5; *Iranian Oil Weekly*, no. 185, February 1975, p. 39.
75 *Moscow Narodny Bank Bulletin*, 9 June 1976, p. 5; *Review of Sino-Soviet Oil*, June 1977, p. 38; Andreasian and Kaziukov, p. 202.
76 *Petroleum Economist*, January 1975, p. 35; *New York Times*, 26 March 1975, p. 63, 10 December 1976, p. D3; *Wall Street Journal*, 26 March 1975, p. 7.
77 Stern, p. 71.
78 *VT SSSR*, 1918–66, pp. 80–3, 178.
79 CIA Gas, July 1978, p. 22.
80 *Moscow Narodny Bank Bulletin*, 11 January 1978, p. 6.
81 Suzanne F. Porter, 'East-West trade financing, an introductory source', US

Department of Commerce (Washington, DC: US Government Printing Office, September 1976), p. 54.

82 *Moscow Narodny Bank Bulletin*, 9 June 1976, p. 17; Ray, op. cit., p. 39.

83 *Business Week*, 28 July 1975, p. 32; *Foreign Trade*, March 1976, p. 48; *Review of Sino-Soviet Oil*, September 1976, p. 56.

84 Kiichi Saeki, 'Towards Japanese co-operation in Siberian development', *Problems of Communism*, May–June 1972, pp. 7–8.

85 *Review of Sino-Soviet Oil*, January 1979, p. 51; *New York Times*, 7 June 1979, p. 16.

86 *Moscow Narodny Bank Bulletin*, 19 October 1977, p. 5; 12 April 1978, p. 10.

87 *Foreign Trade*, January 1979, p. 53.

88 *Moscow Narodny Bank Bulletin*, 1 February 1978, p. 8.

89 *Literaturnaia gazeta*, 17 December 1975, p. 9; *Soviet Export*, vol. 3, no. 120, 1979, p. 54.

90 CIA Gas, July 1978, p. 55; *Review of Sino-Soviet Oil*, July 1979, p. 51.

91 *Petroleum Economist*, January 1976, p. 5; *Review of Sino-Soviet Oil*, December 1976, p. 51; CIA Gas, 1978, p. 55.

92 Stern, p. 223.

93 *New York Times*, 20 November 1978, p. A-4; *Kommunist*, 8 September 1968, p. 4.

94 *Wall Street Journal*, 18 June 1973, p. 20.

5

The 1973 Energy Market, Its Aftermath, and the Soviet Reaction

OUT-OPECING OPEC

By the early 1970s, the Soviet Union was no longer regarded as a nuisance or a threat by the big seven international oil companies. The traditional pattern of control and distribution of petroleum had been radically recast. Whereas before the accepted procedure had been for each of the seven large oil companies to pump the oil from its wellhead concessions, refine it in its refineries, ship it in its tankers and pipelines, and finally sell it to its gas pumps, now there were any number of variations on that theme. Indeed, it was harder and harder to find companies that were able to maintain such all-encompassing integrated networks operating from a foreign base. More and more of the countries where petroleum was produced had nationalized their fields. In addition, various new and independent private and national oil companies had been created all over the world. Some produced oil, some shipped it, some refined it, and some retailed it. In other words, the Soviet model of operation where a state agency served variously as an exporter, shipper, refiner, wholesaler, and retailer, at one time an anomaly, was becoming more and more common. Even though in the early 1970s it had become the world's third largest producer after the United States and Saudi Arabia, the Soviet Union was no longer thought of as a disruptive element. It had oil for export at a time of growing market tightness. Instead of being treated as a pariah, by the early 1970s the Soviet Union was being treated more and more as a partner.

With its new status, the USSR was ideally situated to take advantage of the oil embargo that was declared shortly after the start of the Yom Kippur War; nor did the embargo take the Soviets by surprise. They did all they could to induce the Arab oil producing countries to set up that embargo. Prior to and during the war, the Soviets increased not only the flow of arms but also economic advice. The main thrust of the latter was that if war broke out the Arabs should withhold petroleum from the non-communist Western and Japanese world.[1] Such an embargo, the Soviets argued, would cause a 'major commotion in all the countries of the capitalist world'.[2] The Soviets kept up their call throughout the summer and fall. Radio Moscow on 25 September 1973, a few days before the invasion, urged the Arabs 'to use oil as a weapon against Israeli aggression', and four days later by extension to 'use oil as a political weapon against imperialism'.[3] Then, once the 'weapon' was actually used, the Soviets insisted in the strongest way that the Arabs should not back down and end the embargo. Thus Radio Moscow in its Arabic broadcast warned: 'If today some Arab leaders are ready to surrender in the face of American pressure and lift the ban on oil ... they are taking a chance by challenging the whole Arab world and the progressive forces of the whole world, which insist on the continued use of the oil weapon.'[4] Such exhortations continued into March 1974, even though the fighting had already ceased, and OAPEC (Organization of Arab Petroleum Exporting Countries) a few days later would end the embargo.[5]

The Soviet desire to see the embargo prolonged was due not only to its political impact. Undoubtedly the longer the embargo continued, the more it would hurt the non-OPEC nations, especially the United States, which along with the Netherlands was subject to the most severe restrictions. The Soviets also had an economic interest in what was happening. Along with other non-OAPEC countries, such as Indonesia, Iran, and Nigeria, the Soviet Union found itself with a unique market opportunity. The market price of oil had soared fourfold. Moreover almost all the importers, not only the United States and the Netherlands, found their supplies cut by 10 to 20 per cent.[6] Refusing to let principle stand in the way of profit, the Soviets did their best to take advantage of the situation. Petroleum exports in 1973 shot up by 11 million tons, the largest jump in Soviet history. Although in 1971 they had jumped 9 million tons, the normal increase was about 5 to 6 million tons a year and in 1972 exports had increased only by 2 million tons.

More than that, Soviet exports rose to the United States and the Netherlands. That Soviet petroleum was shipped during the embargo is indicated by the sharp increase in the price the Dutch had to pay for their 1973 oil. Whereas they paid only about $13 a ton for petroleum in 1972, in 1973 the price was $42 a ton, three times as much. In the case of the United States, American trade figures indicate that the Soviets exported some $40 million worth of petroleum during the last quarter of 1973, that is, at the height of the embargo.[7]

Since the Soviet Union was not a member of OPEC, much less of OAPEC, there was no official reason why it could not sell to anyone it pleased. Yet because they had put so much pressure on OAPEC to embargo, the Soviets were highly sensitive to any charges that they were circumventing the embargo.[8] At first they denied any duplicity and expelled a Swedish reporter who had the audacity to report such rumors.

Nevertheless, not only were the Soviets charged with violating the embargo, they were also accused of profiteering at Arab expense. When the Arabs withheld exports from the market, the Soviets rushed in to charge what the market would bear. Given the panic set off by the embargo, it bore a lot. Nor did the Soviets limit their price increases to just the Dutch and the Americans. Neither were they restrained by the fact that their original contracts called for delivery at about $2 to $3 a barrel. In March 1974 the Finns agreed to pay $17 a barrel to the Soviets, an eightfold increase in price. The Soviets were even reported to have asked an Italian company and a German company for $18 a barrel, but at that point, apparently, there were no takers.[9] Soviet hard-currency earnings in 1973 from the sale of petroleum increased by $600 million, or double the 1972 earnings.

Such charges about profiteering and violating the embargo were particularly embarrassing because the Arabs had excluded the Soviet Union from the embargo. Iraq in particular had exported over 11 million tons of petroleum to the Soviet Union in 1973. However, the Arabs had good reason to suspect that once it was consigned to the USSR the latter secretly contracted to resell that petroleum on the spot market where at least some of it could have ended up in the United States and the Netherlands.[10] Rumania was said to have engaged in the same practice. The Iraqis were incensed at this deception and cut back their exports to the Soviet Union the following year to 3·9 million tons. (See Table 4.4.)

The Soviets were hard-pressed to defend themselves. At first they denied that there were any plans to increase deliveries or to increase prices.[11] Ultimately they acknowledged that they had resold Arab petroleum, but there was nothing to be upset about, since it was only diverted 'with [Arab] consent to other Socialist states'.[12] What the Soviets did not explain, of course, was that even if Arab oil did not go directly to Western countries such as the United States or the Netherlands, its diversion to Eastern Europe none the less freed Soviet petroleum that could and did go to the West.

Soviet cynicism toward the oil embargo was further evidenced by the Soviet decision to go ahead with the inauguration of its gas pipeline to West Germany. By chance the pipeline was scheduled for opening 1 October 1973, right at the start of the Yom Kippur War. Rather than suspend shipments and lose the hard-currency exports, the Soviets began the shipment of their natural gas oblivious to all efforts by their Arab allies to create an energy shortage in order to bring pressure on Western Europe.

THE ONE-CROP ECONOMY

Overall the oil embargo and the fourfold price increase was a bonanza for the Soviet Union. Not only did hard-currency earnings for petroleum exports double in 1973, they also doubled in 1974 and continued to increase in the years that followed. In a short time, petroleum accounted for 50 per cent of all Soviet hard-currency earnings. As we have seen, petroleum exports since the mid-nineteenth century had played an important role, but never so important. Looking at the Soviet balance of trade in the 1970s, one would quickly conclude that the Soviet Union had become a one-crop economy – petroleum.

Given the fact that the Soviet Union is the world's second largest industrial power and the world's largest exporter of machine tools, one would have thought that in the years after the Second World War machinery exports would gradually come to play an increasingly important role. In fairness, they have. Where machinery exports accounted for only 5 per cent of all the Soviet exports in 1938, they had grown to 12 per cent in 1950, 21 per cent in 1960, and 24 per cent in 1972.[13] Yet all but a small fraction of those exports were destined for the other members of CMEA or the soft-currency developing countries. Virtually none of it is saleable in the OECD countries. Typically, the Soviets import more than twenty times the amount of machinery they export to a country like Germany or Japan, a situation not much different from that which prevailed prior to the revolution. Based on the way OECD importers treat the Soviet Union, the Soviet Union could be considered a developing country and a raw material producing colony. For that matter, raw materials also are an important export earner in Soviet sales to the socialist countries. In 1977, they accounted for over 50 per cent of all Soviet exports to CMEA.[14]

Just how important raw material exports are is most clearly seen in the hard-currency sector. Of course, this is in large part a reflection of the ever-growing role of petroleum. Since petroleum earnings account for 50 per cent of the hard-currency earnings, it should not be surprising that raw materials, as a whole, generate over 80 per cent of all the Soviet hard-currency earnings.* (See Table 1.1.) Even prior to the 1973 price increase, petroleum and raw materials were important. In 1972, for example, petroleum exports generated about 20 per cent

---

* Note that Finland is not included as a hard-currency country. It trades with the Soviet Union on a barter basis. *Foreign Trade*, April 1978, p. 50, September 1977, p. 3. If it were included, petroleum exports would be even more important, since Finland imports so much petroleum. Some Western economists have none the less included Finland in their calculations about hard-currency earnings and thus come up with misleading results. Edward A. Hewett, 'Soviet primary product exports to CMEA and the West', American Association of Geographers, Washington, DC, *Project on Soviet National Resources in the World Economy*, no. 9, May 1979, p. 6.

of all hard-currency exports, still the most important of all the raw materials. Timber, which earned about 14 per cent, was the next most important commodity. All raw materials accounted for 63 per cent of the foreign export hard-currency earnings. Beginning in 1973, petroleum price increases sparked similar increases in other raw material prices. Each year raw materials and petroleum exports increased, so that by 1976 raw materials were responsible for 80 per cent and petroleum 50 per cent of all the Soviet Union's hard-currency earnings.

OIL AND THE BALANCE OF TRADE

For the Soviet planning officials in charge of keeping Soviet exports and imports in balance, the dominating role of petroleum and raw materials in many ways makes their job much simpler. Since there is really only one important export-earner, when the planners find it necessary to increase exports to bring their deficit into balance they know that their problem will be most effectively solved if they can find some more petroleum. Of course, they can also try to increase timber exports, which in recent years have normally been good for about a billion dollars, and natural gas, which earned 1 billion dollars for the first time in 1978. Still, when it is necessary to think in terms of several hundred million or a billion dollars' worth of *new* export earnings, only petroleum can generate that much. Soviet planners know that petroleum commands a market and is highly sought after in the hard-currency world.

Yet even if it made the deficit-reducers' job more complicated, it would be beneficial for the Soviet Union if it were able to export manufactured and more sophisticated goods. This is easier said than done. For years, the Soviets have spoken of the need to increase their export of high-quality manufactured goods and to set up special export-oriented factories, but so far little has materialized.[15] As a result, foreign trade planners have really only petroleum to manipulate when they want to adjust exports in any meaningful way. That appears to be exactly what they have done.

This warrants more discussion because in our consideration of the validity of the CIA report about the energy crisis in the Soviet Union, it is necessary to understand the critical role played by petroleum in Soviet foreign trade. While it is hard to judge precisely what determines Soviet actions, there seems to be good reason to believe that the decision about how much petroleum and other raw materials to export is dependent at least in part on the Soviet need to eliminate or reduce its foreign trade deficit. This balancing process has become particularly important since 1972. Since then the Soviet Union has suffered at least three major

Table 5.1 Soviet Trade Balances with Hard-currency Countries ($m.)

	1971	1972	1973	1974	1975	1976	1977	1978
Capitalist data								
Imports from USSR	2,553	2,915	4,561	6,839	7,166	8,803	10,548	12,387
Exports to USSR	2,251	3,328	4,894	6,258	11,086	12,106	12,112	13,862
Balance for USSR	+302	−413	−333	+581	−3,920	−3,303	−1,564	−1,475
Soviet data								
Exports to West	2,319	2,491	4,327	6,739	6,346	8,420	10,187	11,145
Imports from West	2,429	3,565	5,254	6,910	11,419	12,574	11,845	14,413
Balance	−110	−1,074	−927	−171	−5,073	−4,154	−1,658	−3,268

Exchange rates: 1971 1 ruble = $1·11
1972 1 ruble = $1·21
1973 1 ruble = $1·34
1974 1 ruble = $1·34
1975 1 ruble = $1·32
1976 1 ruble = $1·34
1977 1 ruble = $1·37
1978 1 ruble = $1·50

Table 5.2 Soviet Oil and Natural Gas Export by Class of Country

| | Petroleum (million tons) | | | | | | Natural Gas (billion cubic meters) | | | | |
| | Output | Export | | | Crude | Refined | Output | Export | | | Import |
		Total	CMEA	Hard Currency				Total	CMEA	Hard Currency	
1972	400	107	57	34	76	31	221	5	3	2	11
1973	429	118	63	36	85	33	236	7	4	2	11
1974	459	116	67	31	80	36	261	14	5	5	12
1975	491	130	72	38	93	37	289	19	11	7	12
1976	520	149	78	49	111	38	321	26	13	11	12
1977	546	155–159	81–85	51			346	32–34	15–16	17	12

harvest short-falls. In addition, the Soviet Union has been whipsawed by first the surge in raw material prices and then the world recession. Through it all, petroleum has served as a balancing mechanism that has prevented the trade deficit from becoming even more serious. Moreover, at times petroleum was used for this purpose, even though by doing so it occasionally had a serious effect on the domestic economy.

Since Gosplan, Gosbank, and Ministry of Foreign Trade officials tend to be very secretive about how they determine their export policy, some notion of what took place and how petroleum was used as a balancing mechanism can be surmised from an analysis of Soviet statistics. For this purpose it is useful to compare Table 5.1, Table 5.2 and Table 4.2 *

Note Soviet export behavior in 1973. Soviet petroleum exports to the hard-currency world increased by 2 million tons.[16] Combined with the price increase during the last quarter of 1973, this was enough to cause a doubling in hard-currency oil revenues. Given the fact that the Soviets also found themselves in 1973 with a grain import bill of $1·1 billion, without the $700 million or more in additional petroleum exports, the deficit would have caused enormous distress at a time when trade deficits totaling $1·5 to $2 billion were most unusual, if not cause for economic alarm.

The abrupt reversal in the terms of trade, which carried over into 1974, seemed to herald a new age. Raw material producers were no longer the underdogs. Economists and political observers were predicting OPEC-like cartels in several other products as well.[17] The rejoicing was shared by the Soviet Union. If it could not attain economic power through the mastery of sophisticated technology, then it would do so via the unexpected earning power suddenly made possible by the export of raw materials. Soviet export earnings soared by about $2·3

---

* The careful reader will note that there are some discrepancies in the tables that are not entirely explained by the process of rounding-off or the usual statistical explanations. In Table 5.1, for example, strange as it seems, export and import figures collected by the International Monetary Fund (IMF) from capitalist sources usually show the Soviet Union has a lower trade deficit than a similar comparison made with Soviet data. The reason for that is that the Soviets do not include the export of various precious metals, including gold, in their export figures. Barry Kostinsky, *Description and Analysis of Soviet Foreign Trade Statistics*, Foreign Demographic Analysis Division, Bureau of Economic Analysis, Social and Economic Statistics Administration, US Department of Commerce, Foreign Economic Reports, No. 5, 1974, p. 53. As a result, Soviet export figures are consistently $100 to $800 million below those reported by the West. Similarly the calculation of the ruble value of hard-currency petroleum exports presented in Table 4.2 does not always equal the overall dollar figure presented in Table 1.1, because the latter table sometimes includes Western statistical information omitted by the Soviets.

billion. As a result, according to the IMF data, the Soviets accumulated one of the largest hard-currency surpluses in their history. Some Western bankers even began to talk about the prospects for ruble convertibility.

The 1974 trade balance was helped by the fact that 1973 and 1974 were good harvest years. Soviet grain imports fell by about $500 million from 1973. But the most important change was in petroleum export earnings. Hard-currency earnings from petroleum sales doubled in 1974 to $2·6 billion. Even more to the point, this doubling took place despite the fact that petroleum exports to the hard-currency countries actually *fell* by 5 million tons. With earnings at such a high level and with the resulting trade surplus (according to the IMF) or manageable deficit (according to Soviet statistics), there was no need to export more. The Soviets have never been taken with the mercantilist credo of exporting for export's sake. Normally the Soviets only export because they must pay for imports. Once they earn enough to pay those bills, that is usually all they will do. Certainly that seemed to be the case in 1974.

Yet like other raw material producers at the time, the Soviets apparently predicated their future trade plans on the assumption that this new favorable balance in the terms of trade was not just a temporary condition. Therefore, like their fellow-beneficiaries in OPEC, they charted for themselves a robust expansion of foreign trade imports. The Soviets all but doubled their purchases in 1975 over 1974 from $6·3 billion to $11·1 billion. In part this reflected price inflation in manufactured goods, and in part the bad harvest in 1975, which had led to an increase in hard-currency expenditures for grain of about $1·5 billion. But more than anything else, the record size of their imports reflected the expectations that all raw material producers in addition to OPEC felt at the time.

Inevitably the rise in petroleum and other raw material prices was followed by a worldwide recession. But unlike some of the other raw material producers, the Soviets failed to retrench in time. In large part this was because they had fallen victim to their own ideology. For years, Soviet economists and ideologists had preached that their economy was immune to the ups and downs affecting the capitalist world. Not only was the Soviet ruble insulated from foreign speculators and manipulators, they argued, but so was Soviet production and employment. Since past Soviet employment and production did indeed appear to follow a course independent of the business cycle swings in the United States and Western Europe, their analysis seemed to be correct. But that was when the Soviet economy was much more insular. In the interim, foreign trade or at least some aspects of foreign trade had become more important to the Soviet economy. Foreign exports as a percentage of gross industrial product had risen from roughly 2–3 per cent of the gross industrial product to about 7 per cent in 1978.[18] Generally, even 7 per cent is so small that foreign fluctuations do not have much impact on domestic affairs. What was ignored, however, was that the interaction of certain segments of the Soviet economy with foreign markets was much higher. Thus, in the extreme case of petroleum, as we saw, the Soviet Union exports 27 per cent (allowing for imports) of its production. Exports of other commodities were not as high a percentage of production and not as much of the product ended up on the capitalist markets. Nevertheless, the Soviets do export over 10 per cent of their manganese, iron ore, glass, and cotton. Equally important, in a bad year like 1975 they had to import an amount equal to 11 per cent of their grain crop (about 80 per cent of which came from the hard-currency world).

Still ignoring all of this, the Soviets continued to import on a business cycle as usual basis. Only belatedly did they discover that they were not as immune as they had talked themselves into believing. They forgot that, despite what the Soviet Union might do, when a recession hits the capitalist world the effect of income elasticity on the demand for raw materials causes an instant drop in the demand for raw materials. Thus, while Soviet imports in 1975 increased by over $4·5 billion, most other countries cut back sharply. Reflecting this contraction, Soviet exports, according to IMF data, increased by only $300 million and according to Soviet statistics, Soviet exports actually decreased by $400 million. The other countries of the world simply were not buying as much. Consequently, within a year's time, the Soviet record trade surplus had turned into a record trade deficit. The 1975 poor grain crop plus the recession proved too much for the Soviet economy. As Soviet planners found out, they must heed not only Mother Nature, but also Father Capitalism.

It is particularly instructive to see how the collapse of foreign markets affected the export of Soviet commodities. For example, timber sales, which totaled about $1 billion in 1974, fell to $700 million in 1975. Similarly sales of cotton fell from $360 million to $274 million. The demand for energy products remained relatively strong, however, even though prices dropped a bit. As a result the Soviets were able to offset their shortfall in the other markets with an increase in the absolute volume of energy products sold in the hard-currency markets. Hard-currency coal revenues rose from $230 million in 1974 to $371 million in 1975. Natural gas exports to the hard-currency world rose from 5 billion cubic meters to 7 billion cubic meters, and more important, revenue more than doubled from $87 million to about $200 million in 1975. However, increased petroleum sales provided the biggest supplement. Hard-currency exports rose from 31 to 38 million tons, and earnings rose from $2·6 billion to $3 billion. However, this apparently was less of an increase than the Soviets had anticipated. This is indicated by an analysis of the data which shows that the Soviets had to settle for prices that were 5 to 6 per cent lower than they had charged in 1974. In addition, the Soviets found themselves with a year-end inventory of fuel that was almost double the size of any past inventory, and one which they promptly reduced once export markets improved.[19] Even so, without this extra petroleum to export, Soviet exports in 1975 would have been $400 to $500 million less than they were in 1974.

It would appear that the Soviets' need for grain and the Americans' need for petroleum were complementary. Why not arrange a swap? It seemed like a perfect match. For a time, during the fall of 1975, American and Soviet negotiators, engaged in serious talk about such a deal, seemed to be close to an arrangement. The American negotiators, however, sought to obtain price concessions from the Soviet side. The Soviets were unwilling to make such concessions, concluding that a poor harvest would not be an annual affair and, more important, that the shortage of petroleum would be with us for some time. Without the price concessions, the American side lost interest. In principle, it was opposed to the making of such bilateral oil deals. After a few months, the talks terminated. Still the Soviets were badly in need of grain, particularly after the American government, fearing another massive Soviet raid on American grain stocks, declared an embargo on American grain shipments in August 1975. Eventually, under the pressure of the embargo, the Soviets agreed to sign a five-year American–Soviet grain pact. According to the new agreement, the Soviets promised to buy a minimum of 6 to 8 million tons of grain a year. They could buy more, but only with special permission. Nothing was said about petroleum.

THE CONSERVATIVE RESPONSE

The record Soviet 1975 trade deficit and the mounting concern by foreign banks about Soviet liquidity did not pass unnoticed. Jarred by the threat that it might lose its credit standing, the Soviet Union reacted as if it had received a visit from its friendly banker at the IMF – it cut back on imports and increased exports.

Decreasing imports was not easy. Although the 1976 harvest was a record one, it took until the end of the year before the normal flow of domestic grain was large enough to substitute for foreign supplies. Thus during 1976 the Soviet Union had to spend $2·5 billion on grain imports, over $300 million more than in 1975. Despite this need for extra money for grain and the ever-rising costs of industrial projects already under way, the Soviets were able to hold down the overall increase in their import expenditures to about $1 billion compared to an increase of $4·5 billion the preceding year.

Although virtually the entire burden fell on Soiuznefteexport, increasing exports was considerably easier. As the world economy began to recover, demand and prices for raw materials also firmed. Thus, hard-currency earnings of timber and cotton each increased by about $100 million. This contributed to the $1·6 billion to $2 billion increase in Soviet export earnings, but the effect is small by comparison with the $1·5 billion increase in hard-currency petroleum earnings. In determined fashion, the Soviets increased petroleum exports by 11 billion tons just to the hard-currency world. (See Table 5.2.) This was an unprecedented amount. Except for 1973 and 1975, the combined increase of petroleum exports to *both* hard- and soft-currency countries had never totaled that much. As it was, the overall increase in exports in 1976 amounted to 19 million tons. That was equal to two-thirds of the total increment in petroleum production for that year. Normally no more than 40 per cent of a year's increase in production is set aside for export.

Senior officials at Gosplan in an interview in December 1978 acknowledged that this surge in hard-currency exports had been deliberate. As N. N. Inozemtsev, the deputy chairman, and his deputies explained, the decision to divert such a large percentage of production in order to reduce the trade deficit was taken in 1976 by 'Gosplan in consultation with the Ministry of Foreign Trade and Gosbank and sent to the Government'. While Inozemtsev's answer will win no award for precision or detail, it does indicate that this major policy change was a specific response to an emergency situation. Even though it involved a $1·5 billion transaction, it was not part of any long-term planning process.

As a result of this, and continued belt tightening in 1977, the Soviets were able to make substantial reductions in their trade deficit. Whereas the deficit total was $4 to $5 billion in 1975, and $4·1 to $3·3 billion in 1976, it was reduced to about $1·5 billion in 1977. In part this was made possible by the good 1976 harvest and the relatively good one in 1977. This made possible a reduction of over $1 billion in hard-currency grain imports. According to Soviet data, this in turn made possible a reduction in overall imports by as much as $700 million. (See Table 5.1.) As in the years previously, enhanced petroleum earnings of $800 million generated the bulk of the $1·7 billion increase in hard-currency income. However, because of a 10 per cent OPEC price increase in early 1977, these increased earnings were due more to the higher charge per ton than to a major increase in export tonnage.[20] In

1977 the Soviets apparently only increased their hard-currency exports by about 2 million tons.*

EASTERN EUROPE

Such tumultuous developments in the non-communist world inevitably affected the CMEA bloc as well. Yet, it was remarkable that initially the impact was relatively minor. Many of the East European countries increased their retail gasoline prices.[21] Still none of them was embargoed by OAPEC. In any event, most of them imported little petroleum from the Middle East. All but a small percentage of their petroleum came from the Soviet Union. Rumania, of course, was different. It had been trading with Israel and using the Israeli pipeline from Elat to Ashkelon to obtain its oil from Iran.[22] Some members of OAPEC were not exactly overjoyed about this, but since Rumania was a net exporter of petroleum until 1976 Rumania did not overly worry about OAPEC pressure.

While the East Europeans as a whole may not have been greatly affected by the actions and possible actions of OAPEC, they were very much concerned about the Soviet Union. Would the Soviet Union cut back its exports to Eastern Europe or raise its prices on petroleum? According to a procedure agreed to by CMEA in 1958, the 1971–5 prices for intra-CMEA trade were based on the world price of petroleum for the preceding five years, 1966–1970.[23]

In the early 1970s, this seemed to be a reasonable enough procedure. After all, the price level for raw materials like petroleum had been relatively stable. Moreover, since the Soviets always managed to add something extra to the price of petroleum exported to Eastern Europe, they felt themselves well enough protected. It was a shock, therefore, to discover that under the existing arrangement the Soviet Union by 1974 was delivering oil to Eastern Europe at prices as little as one-quarter or one-fifth of the world market level. Moreover, petroleum was being paid for in East European soft currencies. That was all well and good when prices to Eastern Europe were double those which the Soviet Union could earn in the spot market. But in 1973 the Soviet Union was sometimes collecting as much as $17 or $18 a barrel in the world market, but only about $3 a barrel for its sales to Eastern Europe.

---

* Unfortunately, in their foreign trade handbooks issued in 1978 and the years following, the Soviets, in violation of the Helsinki Agreement, which called for the release of more not less economic data, began withholding data showing the actual tonnage of petroleum exports and imports. They did, however, publish the ruble value of such trade. Based on data published elsewhere, it is possible to reconstruct with reasonable accuracy at least the data for 1977. (See Table 4.2.)

This was a topsy-turvy world. Mendershausen and Holzman had argued about why the Soviet Union charged its East European allies relatively high prices, but beginning in 1973 and 1974, it was the capitalist world that was paying a higher price. The East Europeans for all intents and purposes were being subsidized. Obviously, this is not what the Soviets had in mind when the five-year pricing plan went into effect in 1971. More than that, there was now an opportunity cost to selling to Eastern Europe. If all Soviet exports of petroleum in 1973 had been diverted to Western Europe, the Soviets would have earned an extra $4·5 billion to $7·5 billion in hard currency. That was more than all their other hard-currency export earnings.

Yet in 1973 and 1974 the Soviets held to their agreement as to both quantity exported and price. Exports to CMEA went up by 6 million tons in 1973 (at an opportunity cost of over $400 million), and by 4 million tons in 1974 (at an opportunity cost of about $300 million). Moreover, the East Europeans were the only ones in the world able to buy 1974 petroleum at pre-1973 prices.

Finally, it was too much for the Soviet Union to bear. There was no longer any doubt as to who was exploiting whom. In early 1975, a year in advance and in violation of its contractual agreement, the Soviet Union unilaterally forced a price increase on its allies. Henceforth, prices would no longer be set for five years on the basis of the world prices recorded during the preceding Five-Year Plan. Instead, one year's prices would be determined by the rolling average of prices of the immediately preceding period. For 1975, the system was changed further, so that it encompassed only the immediately preceding three years, not five years as would normally be the case. This allowed the Soviet Union to take advantage of the high prices of 1974 and late 1973, without having the effect washed out by the lower prices of 1970 and 1971. This permitted a price increase of about 140 per cent over the 1974 prices, though actual price increases turned out to be lower than that.

In 1976 and thereafter the system reverted to a five-year average, but it was a moving five-year average. Extending the base this way, however, meant that the price increases of 1973 to 1975 were tempered by the lower prices of 1971–2. As a result, the prices of 1976 increased by only about 4 per cent over 1975. Since in 1977 the base encompassed 1972–6, prices to Eastern Europe went up again by about 25 per cent. One-fifth of the 1977 OPEC price increase of 10 per cent was also built into the 1978 price change.[24] This caused prices to go up another 25–30 per cent, at which point they were about four times higher than they were in 1972. Thus in 1979 Hungary was paying 68–70 rubles a ton for Soviet petroleum compared to 16 rubles in 1974.[25] Oil prices to Eastern Europe were still about 20–25 per cent below world prices, but the difference was narrowing.[26]

The 50 per cent or more increase in OPEC prices in 1979 again increased the disparity between what the Soviet Union earned in the world market and what it earned in Eastern Europe. The Soviets were distressed, and there was talk that they would again reduce the rolling average from five to three years or even to one year and base the price charged for petroleum this year to Eastern Europe on last year's world price.[27]

Whatever the absurdity of charging 1972 prices in 1975, the Soviet Union's decision to force through the change a year earlier was technically a violation of a contractual agreement. The East European countries were not happy about the unilaterally imposed change. Never mind that they were still being subsidized as preferential customers. Never mind that, for the sake of political relations within the bloc, the Soviets had honored their contracts and held down their prices as long as they had. The change increased costs in Eastern Europe to the advantage of the Soviet Union, and it complicated East European five-year plans. While it was true that under the new formula the East Europeans could also raise the prices of their exports to the Soviet Union, most East European exports consisted primarily of machinery products whose world prices initially lagged behind the increases that took place in commodity prices. Thus, the East Europeans began to complain about the rapaciousness of their suppliers, even though in the suppliers' eyes the prices were still too low.

It is worth inserting a special note here about the Soviet practice of periodically violating a contract. Generally the Soviets are regarded as faithful observers of contractual agreements. In the overwhelming majority of instances, they are. Yet, as this study has shown, there have been several occasions where this has not been the case, as when the Soviets cut off petroleum exports to Yugoslavia, Finland, China, and Israel. With respect to price they have pushed up their prices despite contractual agreements not only to Eastern Europe in 1975, but in 1973 to some West European buyers, and in 1977 during the dual-price period when they insisted on the higher OPEC price rather than the lower Saudi Arabia base that was stipulated in some of their contracts. If the political or economic forces suddenly change, the Soviets, despite the prevailing stereotype, will not hesitate to disregard their previous commitments.

There is no denying that the 1975 contract abrogation hurt Eastern Europe. Petroleum imports started to eat up a larger chunk of the CMEA countries' import bills. This showed up strikingly in 1975. In 1974, with the old low 1973 prices, 19 per cent of the ruble value of Soviet exports to the socialist countries consisted of petroleum, energy, and electricity, while machinery accounted for 28 per cent. In 1975, after the price increase, energy export earnings increased to 26 per cent of the total, and machinery had fallen to 23 per cent.[28] The impact on the Soviet balance of trade with Eastern Europe was even more striking. Whereas the Soviets had a trade deficit with all the East Europeans of about 700 million rubles in 1973, that gradually turned into a surplus, so that by 1976 the Soviets had a surplus of about 900 million rubles. (Table 5.3.) It peaked at 1·4 billion rubles in 1977. Only Rumania, which imported no Soviet oil, had a trade surplus during this period. As the East Europeans began to adjust to higher petroleum prices, the overall Soviet surplus with Eastern Europe fell sharply to about 170 million rubles in 1978, which probably explains the reports of renewed Soviet pressure to increase prices to Eastern Europe in late 1978.[29]

Table 5.3   *Soviet Balance of Trade with Eastern Europe (million rubles)*

	Export	Import	Balance
1973	7,380	8,093	− 712
1974	8,707	8,600	+ 107
1975	11,867	11,312	+ 555
1976	13,107	12,228	+ 879
1977	15,267	13,852	+ 1,415
1978	16,945	16,776	+ 169

There is also evidence that indicates the Soviets have tried to cut back their petroleum shipments to Eastern Europe. The Czechs and the Hungarians in particular have complained about this.[30] The CIA analyst John R. Haberstroh claims that the Soviets told the East Europeans in the early 1970s that, unless the East Europeans helped the Soviets with their investment program, their 1976–80 oil imports would be held to 1975 levels.[31] This is hardly a new threat since, as we saw, and as Haberstroh elsewhere concedes, the Soviet Union as early as the mid-1960s committed itself to increased deliveries in exchange for East European investment help, and indeed petroleum deliveries did increase. None the less, there is no doubt that the Soviets have become less forthcoming than they were earlier. Among other measures, they have started to insist that any increase in deliveries beyond the limits specified in the plan be paid for in dollars. Since the East Europeans are so short of currency, this is undoubtedly a heavy burden for most of them, but in a few cases it is not. For example, Hungary agreed to pay hard currency, but on the condition that the Soviet Union pay hard currency for the equally fundable Hungarian commodities, grain and meat. By the time the swap arrangements have been completed, the

Soviets traditionally end up owing hard currency to the Hungarians.[32]

Despite all these pressures and despite the fact that the CMEA countries were unable to buy all the petroleum they wanted, the fact remains that their petroleum imports from the Soviet Union since 1972 have increased at a respectable rate. As Table 5.2 indicates, the Soviets have increased their exports to CMEA countries by about 4 million tons a year. That reflects at least a 6 per cent increase for the years for which we have relatively precise data. Overall, Soviet exports rose from 63 million tons in 1973 to 81 to 85 million tons in 1977, a 29 to 35 per cent increase. Given that public consumption in most of the rest of the world actually fell, at least in 1974 and 1975, it appears that the CMEA members did relatively well by Soviet exports. It is unclear how long the Soviets will continue this pace.

Heeding Soviet warnings, the East Europeans have actively sought alternative sources of supply. Even without Soviet pushing, the Middle East was an early target of their economic and military aid. Iraq, Iran, Libya, and Syria in particular were recipients of large quantities of East European foreign aid.[33] Unfortunately, except for Rumania, the other East European countries did not have much to show for all their effort. Table 5.4 shows how much the East Europeans imported from non-Soviet sources (basically the Middle East) in the 1970s. Total imports seldom exceeded 10 million tons a year or 200,000 barrels a day. Except for Rumania, only the East Germans have imported as much as 3 million tons in any year from a source other than the Soviet Union. Even with this supplement, East Germany continues to depend on the Soviet Union for as much as 88 per cent of its imported crude oil.[34] Poland appears to be the most independent, relying on the Soviet Union for about 70 per cent of both its imported crude and refined product. However, this independence may be deceiving. In its effort to diversify its sources of supply, Poland has arranged with British Petroleum (BP) to build a refinery to supply it with up to 3 million tons a year.[35] When this agreement was made, the assumption was that most of this would come from the Middle East. But on occasion British Petroleum also swaps petroleum with the Soviet Union. It is probable, therefore, that at least some of the oil processed by British Petroleum for the Poles was originally Soviet oil.

Even though it only began to import crude oil in 1968, and until 1976 it was a net exporter of petroleum, Rumania by itself often imports more Middle East petroleum than all the other members of CMEA combined. (See Table 5.4.) This is largely because the Rumanians have excess refining capacity and export relatively large quantities of refined petroleum products to other countries. While it obtains these imports from several sources, for a long time its chief supplier was Iran. Thus, the Rumanians were affected by the Shah's

Table 5.4   *Non-Soviet Imports of Crude Oil (thousands of metric tons)*

	1970	1971	1972	1973	1974	1975	1976
Bulgaria	937	105	1,914	2,139	1,620	598	817
Czechoslovakia	398	837	664	1,136	364	336	766
East Germany	1,111	1,165	3,645	3,020	2,299	1,900	2,000
Hungary	398	n.a.	n.a.	792	n.a.	1,474	n.a.
Poland	0	0	0	570	827	2,424	n.a.
Subtotal	2,842	2,107*	6,223*	7,657	5,110*	6,732*	3,583**
Rumania	2,291	2,858	2,873	4,143	4,538	5,088	8,475

* Without Hungarian imports.
** Without Hungarian or Polish imports.

overthrow as much as any country in the world, particularly since its trading terms with Iran were quite reasonable.

In an effort to increase their volume to something like that of Rumania, Czechoslovakia and Hungary have joined with Yugoslavia to build a pipeline from the Yugoslav port of Rijeka on the Adriatic coast to Hungary and Czechoslovakia.[36] The Yugoslavs are to take 34 million tons of petroleum a year (almost 700,000 barrels a day) from the pipeline and the Hungarians and Czechs are to take 5 million tons each (100,000 barrels a day).[37] Rumania also hopes the pipeline will be extended to its borders where it could draw off 3·5 million tons.

No definite statement has been made as to exactly where the supplies for the pipeline will come from, but since Kuwait with $125 million and Libya with $75 million, along with the World Bank, are the main outside investers in the project, it is reasonable to assume that some of the petroleum will be theirs.[38] Presumably, they would ship oil if for no other reason than to assure the return on their pipeline investment. The Iranians, at least until the Shah's overthrow, had also been expected to use the pipeline as well.[39] Strangely enough, there were also reports that the Soviets had plans to use the pipeline. Apparently they expect to move some of the Middle East oil (probably from Iraq) they buy through this pipeline to their East European customers.[40]

Even with all these measures, the opinion is widely shared that the East Europeans will soon find themselves short of oil. Haberstroh shows that the East Europeans had placed significantly larger orders in the Middle East, but they cut back their orders as prices continued to increase.[41] Moreover, they are affected by the cessation of oil exports from Iran. Extrapolating several of these trends, the University of Texas economist Edward A. Hewett calculates that the East European energy deficit will grow from 1975 to 1980 by 60 per cent (or from a deficit of 95 MTSF, million tons in standard fuel equivalent, in 1975, to 152 MTSF by 1980).[42] This is predicted on his assumption that East European energy production will grow at 2·5 per cent a year from 1976 to 1980, while consumption will increase by 4 per cent a year. Hewett questions whether the Soviet Union will be able to satisfy this growing shortfall, presumably with petroleum, and still export to the hard-currency market. He calculates that the Soviets will have an energy surplus in 1980 of about 211 MTSF. That would leave a bloc surplus of only about 60 MTSF, not the approximately 105 MTSF that existed in 1975.[43] The presumption is that the surplus would diminish even more in subsequent years. The bloc as a whole might soon become net importers.

While Hewett's analysis may be correct, there are some offsetting factors. The first is that the ratio of energy consumed to GNP seems to have fallen slightly in the Soviet Union just as it has in the United States.[44] Like the rest of the world, the Soviets are striving to conserve on energy consumption. The same reduction in energy consumption as a percentage of GNP may be occurring in Eastern Europe. Secondly, while Hewett in his analysis properly focuses on the shortfalls the Soviets seem to be suffering in coal and oil production, he neglects the success the Soviets seem to be having in gas production and in the substitution of gas for oil. Not only is gas becoming more important within the USSR, it is also becoming increasingly important in Eastern Europe now that the Orenburg–Soiuz pipeline has opened. In 1979, the planned flow of the pipeline was 8 to 9 billion cubic meters of gas to Eastern Europe, and in 1980 its planned flow was to be 15·5 billion cubic meters.[45] That would provide 2·8 billion cubic meters of gas a year for Bulgaria, Czechoslovakia, East Germany, Hungary, and Poland. Rumania will receive 1·5 billion cubic meters. That is the equivalent of about 2·4 million tons of oil, or about 14 per cent of the petroleum Czechoslovakia or 25 per cent of the petroleum Hungary imported from the USSR in 1977.

Finally, there is a chance that Czechoslovakia may yet receive some of the 2 to 3 billion cubic meters of gas a year it was scheduled to be sent in the complicated, and for the time aborted, Iranian–Soviet Union–Western Europe natural gas deal. The Soviet Union may decide to carry through on its part of the contract and collect the hard currency for itself regardless of what the Iranians decide. It has the gas and the pipeline capacity in the Orenburg pipeline. If it can find enough gas to divert to its own south, where the Iranian gas would have been used, it would make sense for it to move ahead on its own. For that matter, the British analyst Jonathan Stern reports that the Soviets may be obligated to ship their gas by the terms of the contract. Apparently the contract anticipated that there might be a problem in Iran and therefore requires performance by each party regardless of what happens to another party.[46]

The significance of these gas shipments is that they will reduce the energy deficit in Eastern Europe as calculated by Hewett from 152 MTSF to 130 MTSF. More important, the deficit will be reduced, not with petroleum, but with the shipment of natural gas, which for the most part has already been set aside. This reduction should be taken into account before any supplemental amounts of natural gas are added and before the East Europeans and Soviets have to haggle over increasing the allotment of oil and coal exports. Of course this increase in natural gas output in the Soviet Union is included in the calculation of Soviet energy surplus. But the significance here is that this is above and beyond any petroleum exports and before the Soviets begin to provide for their own needs.

Although it is not of the same magnitude, Hewett apparently also neglects to allow for the Soviet shipment of electricity in the East European electrical grid. In 1978, the Soviets exported 12 billion kilowatts of electricity or 1 per cent of their total production. Except for Finland, all of this went to Eastern Europe. With the opening of a new transmission line in 1978, the Soviets expect to supply another 6·4 million kilowatts and more later.[47] The first stage alone should add another 2 MTSF. Just as with natural gas, the increased export of electricity will substitute for petroleum.

Nuclear energy is another substitute for Soviet petroleum that Hewett acknowledges but apparently also omits from his calculations. While the nuclear energy construction program has not always been on schedule, the East Europeans, just like the Soviets, only have to contend with the confusion of their planning and production system, not the opposition of environmental and regulatory agencies. In the case of Eastern Europe (we shall consider the Soviet Union in Chapter 7) this means that sooner or later they will complete production on a whole series of projects in each country. The Bulgarians expect that by 1980, 20 per cent of their electric energy will be generated by nuclear power. This will give them 1,760 megawatts (MW) of nuclear capacity.[48] In the case of East Germany, the hope is that by the year 2000 nuclear power will constitute 50 per cent of its electrical power needs.[49] Two plants, generating 5 per cent of the country's nuclear energy, were already operating in 1979.

The Czechs completed their first plant in 1972 and expect to have 350 MW capacity shortly.[50] By 1990, it should be 8,000–10,000 MW. Hungary, Poland, and Rumania (which bought at least one of its nuclear plants, naturally enough, outside the Soviet Union from Canada) have encountered production delays, but expect their facilities to be operative in the early 1980s. Overall electricity generated by nuclear power in Eastern Europe was planned to be 9 per cent of the total, with a planned capacity of 30 million kilowatts of power by 1980.[51]

To supplement these efforts, the Poles, Hungarians, Czechs, and the Soviets in March 1979 entered into a joint venture to build two nuclear energy plants with a generating capacity of 4 million kilowatts on Soviet territory.[52] The East Europeans, just as they did with the Orenburg pipeline, have agreed to finance a substantial portion of these two plants. As repayment, the Soviets will send one-half of the electricity generated in these plants through a 750-kilovolt transmission line from Khmelnitskii to Zheshuv (Rzeszow), Poland.

Many of the East European countries have also concluded that they might have to return to a greater reliance on coal and lignite. Like the United States, Western Europe, and the Soviet Union, the East Europeans in the late 1950s eventually began to move away from greater reliance on coal and lignite to petroleum. But the shift came so late that the readjustment back to coal and lignite may not be all that difficult, at least in terms of production. Ben Korda calculates that as recently as 1974 countries like Poland depended on coal for 94 per cent of their energy. The figures for Czechoslovakia were 76 per cent and East Germany 71 per cent.[53]

While the production facilities may be usable, in many ways this re-emphasis on coal and lignite may be as environmentally hazardous as the decision to move toward nuclear energy. Except for some of the Polish coal, most of the coal available in Eastern Europe is bituminous and is almost as heavy in particulate matter and sulphur content as the lignite. Moreover, it is not only burned by the electric utilities and by industry, but it is used as a source of household fuel. The result is a heavy layer of smog and grit over most East European cities, especially in the winter. Unfortunately, or perhaps fortunately, there is not an enormous amount of untapped mining potential available. But since the decision to move away from coal was a relatively recent one, not too much damage has been done to most of the mining sites. Thus, unlike the case in Western Europe, not too much will have to be spent to refit the mines. When a country like Rumania finds itself short of coal reserves, it may resort to a more ingenious solution. For example, the Rumanians have entered into a joint venture with Occidental Petroleum to share mining costs and coal at their Rock Creek Mine in the United States. They have a similar venture with Quinette Coal Ltd in Canada.[54] Whether the coal comes from the United States or from local East European mines, the decision to maintain reliance on coal and lignite will not be a popular one, but it is one alternative if Soviet gas and petroleum exports prove to be inadequate.

Finally, the East Europeans, regardless of how forthcoming the

Soviet Union may be, can seek to increase the efficiency of their energy use. Ben Korda has found that Czechoslovakia, for example, is so wasteful of its energy reserves that energy consumption per unit of GNP was higher in Czechoslovakia in 1972 than it was in West Germany.[55] Extending his analysis, he and Ivan Moravcik find the same waste plagues all the East European countries as compared with Western Europe.[56] While there are methodological criticisms of this type of analysis, (lignite and brown coal in Eastern Europe tend to be weighted at a higher fuel content than is justified) there is no disputing the fact that fuel consumption in Eastern Europe is inefficient.[57] This is particularly true in industry, where the largest portion of the coal is utilized. However, if they cannot reduce such waste or find substitutes at home or abroad for Soviet petroleum, there is no doubt that, as the CIA and others have predicted, East European economic growth will be seriously affected. Certainly it will be no simple matter to reduce, not to mention eliminate, such waste. At the same time, there is no doubt that much more could be done than is being done to stretch East European energy supplies further.

CONCLUSION

The oil embargo and the fourfold price increase following the Yom Kippur War provided an immense windfall for the Soviet Union. Nor was it a one-time event. As petroleum production increased in the Soviet Union, it found itself leading the world in petroleum production, and by the late 1970s it was second only to Saudi Arabia as the world's leading exporter. It is true that about one-half of those exports went to Eastern Europe, which, of course, did nothing to lessen Eastern Europe's dependence on the Soviet Union, but roughly one-third of all exports, or 50 million tons (1 million barrels a day), were sold for hard currency. As a result, the Soviet Union benefits more from OPEC price increases than most of the actual members of OPEC themselves.

With such large exports, petroleum came to dominate Soviet hard-currency earnings. By the mid-1970s, petroleum constituted over one-half of what the Soviet Union earned in sales to hard-currency countries. It was not that the Soviet Union took delight in exporting petroleum for its own sake. The Soviets would have preferred to export manufactured goods, but since there were few buyers for their manufactured goods they had no choice. Thus petroleum came to be used as the balancer. When the harvest was good and there were no sudden shocks from either recession or inflation, the Soviets tended to hold down the increase or even cut back on petroleum exports. However, when faced with a bad harvest and a big import bill, the Soviets increased their petroleum exports on one occasion by as much as 11 million tons in one year. Since the Soviets have virtually nothing else to replace petroleum exports in this one-crop economy of theirs, the loss of this export commodity, which is both an economic and political weapon, would be a most serious matter for them. We turn now to an examination of just what the energy production prospects are for the future, and what options the Soviets have for solving their balance of foreign trade problems.

NOTES: CHAPTER 5

1　*London Times*, 20 November 1973, p. 18; *New Times*, nos 45–6, November 1973, p. 8, no. 48, November 1973, p. 20; *New York Times*, 13 March 1974, p. 24; Gene Riollot, 'Moscow and the oil crisis', Radio Liberty Dispatch, 10 January 1974; Radio Moscow, 29 September 1973; Radio Moscow, in Arabic, 25 September, 5 November 1973.

2　*Pravda*, 10 June 1973, p. 4.

3　*Jews in Eastern Europe*, vol. VI, May 1974, p. 23; Radio Moscow, 12 September 1973.

4　*New York Times*, 13 March 1974, p. 24.

5　*New York Times*, 13 March 1974, p. 24.

6　Robert B. Stobaugh, 'The oil companies in crisis', in Raymond Vernon (ed.), *The Oil Crisis* (New York: Norton, 1976), p. 179.

7　Bureau of East–West Trade, *Export Administration Report, East–West Trade*, US Department of Commerce, 3rd quarter, Washington, DC, 1973, p. 84; 1st quarter, 1974, p. 60.

8　*Soviet News*, 11 December 1973, p. 524.

9　*Financial Times*, 14 March 1974, p. 24; *Petroleum Economist*, April 1974, pp. 134 and 151; *New York Times*, 15 March 1974, p. 8.

10　*New York Times*, 10 November 1973, p. 4; *The Energy Forum*, The Conference Board, New York, 10 January 1974, p. 4. Iraq was also reported to have sold on the spot market.

11　*Soviet News*, 15 January 1974, p. 15.

12　*Soviet News*, 4 December 1973, p. 508; *New Times*, no. 48, November 1973, p. 22.

13　*VT SSSR*, 1918–66, pp. 17, 73; *VT SSSR*, 1973, p. 19.

14　*VTSR*, 1977, p. 18.

15　*Sotsialisticheskaia industriia*, 2 March 1976, p. 3.

16　See Table 5.2.

17　C. Fred Bergsten, 'The threat is real', *Foreign Policy*, no. 14, Spring 1974, p. 84.

18　Interview at Gosplan in Moscow, December 1978.

19　*Nar. khoz.*, 1977, p. 43.

20　*Petroleum Economist*, July 1977, p. 254.

21　Leslie Dienes, 'Energy policy changes and prospects in East Central Europe', mimeo., p. 17.

22　*Petroleum Economist*, January 1974, p. 13.

23　*Foreign Trade*, November 1978, p. 37; Raimund Dietz, 'Price changes in Soviet trade with other CMEA countries and the rest of the world since 1975', in Joint Economic Committee, US Congress, *Soviet Economy in a Time of Change*, Vol. 1: (Washington, DC US Government Printing Office, 10 October 1979), p. 263.

24　*Review of Sino-Soviet Oil*, January 1978, p. 62.

25　*New York Times*, 16 May 1979, p. A-11.

26 *Wall Street Journal*, 6 July 1979, p. 20; *Petroleum Economist*, March 1979, p. 126; Radio Prague, 15 July 1979.

27 *Foreign Trade*, November 1978, pp. 38–9.

28 *VT SSSR*, 1975, p. 17.

29 *Petroleum Economist*, March 1979, p. 126.

30 ibid., October 1976, p. 393, December 1976, p. 454, November 1977, p. 457; *New York Times*, 15 May 1978, p. D1.

31 John R. Haberstroh, 'Eastern Europe: growing energy problems', in *East European Economics, Post-Helsinki*, Joint Economic Committee, US Congress (Washington, DC: US Government Printing Office, 25 August 1977), p. 379.

32 Interview with Hungarian economist in Moscow, 19 November 1978, and in Boston, 18 September 1979.

33 *Moscow Narodny Bank Bulletin*, 7 January 1976, p. 8, 14 January 1976, p. 12, 27 July 1977, p. 16, 14 June 1978, p. 8, 17 January 1979, p. 14; *Review of Sino-Soviet Oil*, July 1978, p. 61; *Petroleum Press Service*, August 1969, p. 285, June 1973, p. 33.

34 Andreasian and Kaziukov, p. 203.

35 *Moscow Narodny Bank Bulletin*, 12 July 1972, p. 8.

36 *Petroleum Press Service*, June 1973, p. 215.

37 *Review of Sino-Soviet Oil*, February 1978, p. 51.

38 *Petroleum Economist*, December 1975, p. 472, Andreasian and Kaziukov, p. 209, give some different figures.

39 *New York Times*, 29 February 1979, p. D1.

40 *Iran Oil Journal*, no. 181, February 1974, p. 36.

41 Haberstroh, p. 387.

42 Edward A. Hewett, 'The Soviet and East European energy crisis: its dimensions and implications for East–West trade', Center for Energy Studies, University of Texas at Austin, August 1978, mimeo., pp. 22–3.

43 ibid., pp. 22, 24.

44 CIA Gas, August 1978, p. 6.

45 *Review of Sino-Soviet Oil*, August 1978, p. 56.

46 Jonathan P. Stern, *Soviet Natural Gas in the World Economy*, Washington, DC, Association of American Geographers, No. 11, 9 June 1979, p. 43.

47 *New Times*, no. 39, 1978, p. 13.

48 Haberstroh, p. 389.

49 *New York Times*, 15 April 1979, p. 27; *Soviet Export*, vol. 3, no. 120, 1979, p. 28.

50 *Foreign Trade*, December 1978, p. 15; *Current Digest of the Soviet Press*, 20 June 1979, p. 14; *Review of Sino-Soviet Oil*, April 1979, p. 80.

51 *New Times*, no. 39, 1978, p. 12; *Foreign Trade*, February 1977; *Soviet News*, 1 May 1976, p. 193; *Moscow Narodny Bank Bulletin*, 1 August 1979, p. 20; A. M. Nekrasov and M. C. Pervukhin, *Energetika SSSR v 1976–1980 godakh* (Moscow: Energiia, 1977), p. 192; Even after downward revisions in the plans, nuclear energy is still expected to play an important role: *Business Week*, 4 September 1978, p. 47.

52 *Ekonomicheskaia gazeta*, no. 24, June 1979, p. 20.

53 Quoted by Hewett, p. 4.

54 *Moscow Narodny Bank Bulletin*, 18 January 1978, p. 13, 15 August 1979, p. 15.

55 Ben Korda, 'Economic policies and energy production in Czechoslovakia', *ACES Bulletin*, vol. XVII, nos 2–3, Winter 1975, p. 4.

56 Ben Korda and Ivan Moravcik, 'The energy problem in Eastern Europe and the Soviet Union', *Canadian-Slavonic Papers*, March 1976, p. 3.

57 Haberstroh, p. 393.

6

Will the USSR Be Able to Satisfy Its Needs?

From Chapter 5 it should be clear that it is impossible to understand the prospects for the development of Soviet petroleum without some appreciation of the key role petroleum plays in the foreign economic policy of the Soviet Union. Petroleum is the grease that fuels the foreign trade sector. It is also a critical element in Soviet relations with Eastern Europe. Thus, without petroleum, the Soviet Union would face not only political difficulties but enormous economic problems in financing the imports it needs for its domestic industry and agriculture. This is not meant to imply that the Soviets can do the impossible. Sooner or later, they, like everyone else, will run out of petroleum. But before that unavoidable day arrives, there is much the Soviets can do to ensure that they take advantage of all the petroleum reserves potentially available to them. Accordingly they must avail themselves of the latest technology. Similarly they must reduce waste in production and consumption and, if need be, substitute other energy forms in order to divert petroleum for export.

CIA FORECASTS

The CIA has predicted that Soviet petroleum production will level off and then fall sharply. Exports will do the same thing. While the CIA predictions have changed from year to year, and in 1979 apparently did change significantly from some of the earlier dire forecasts in its original and most widely quoted April 1977 report, the CIA predicted in 1977 that in the worst of circumstances Soviet output would peak in 1978 at about 11 MBD (550 million tons) and fall in 1979.[1] In May 1979, after a nine-foot flood in the Tiumin oil fields, production for the whole country dropped back to what it had been in April 1978. Without acknowledging the flood, the CIA pointed to this reduction as an indication that production had leveled off.[2] However, as the flood waters receded, production picked up again in June, so that output continued to grow for the rest of the year, although admittedly at a slower rate. In fairness to the CIA, in its April 1977 report it indicated as an outer limit that output might not peak until 1981 at about 12 MBD (600 million tons) and fall only in 1982. It revised this forecast slightly in 1979 and predicted production would peak at only 590 million tons in 1980.[3] Once output peaks, however, the CIA anticipates a very rapid and sharp drop in production. Thus, by 1985, according to the CIA's most pessimistic scenario, the Soviet Union will be producing only about 8 to 8·3 MBD (400 to 415 million tons). Even in its more optimistic projection, the CIA suggests that production in 1985 will only total about 10 MBD (500 million tons.)[4] This would mean a drastic drop in output of as much as 33 per cent.

EXPORTS OR IMPORTS

Clearly a drop of output to either 400 or 500 million tons from 600 million tons by 1985 would have an enormous effect on Soviet export potential. Domestic consumption of petroleum reached approximately 395 million tons in 1977. (See Table 4.3.) Assuming the petroleum were available, domestic consumption would probably increase by at least 2 or 3 per cent a year. On that assumption, by 1985 consumption of petroleum would range from about 463 to 500 million tons. If production should drop off to 400 million tons, the Soviets would have to import from 63 million tons to 100 million tons of petroleum (1·25 to 2 MBD), depending on whether domestic consumption increased by 2 or 3 per cent. Assuming the best of all CIA worlds, that is, if consumption increased by only 2 per cent but production fell to only 500 million tons, that would leave the Soviet Union only 37 million tons to export to both Eastern Europe and the hard-currency world. To say the least, even this would be a gloomy prospect. Yet it is far better than the more pessimistic CIA prediction that production will be only 400 million tons. With only 400 million tons the Soviet Union would then be forced to import 63 to 100 million tons for itself, plus whatever petroleum was needed in Eastern Europe.

Given the possibility of such a fall in production, it is not surprising that the CIA has predicted that by the mid-1980's the USSR and Eastern Europe will have to import significant quantities of petroleum. In its widely publicized report of April 1977, the CIA put its estimate of imports at 3·5 to 4·5 MBD (175 to 225 million tons a year) for the whole East European and Soviet bloc. After this estimate was criticized by some specialists, the CIA in July 1977 lowered its estimates of 1985 bloc imports to 2·7 MBD.[5] Of that, 1·6 MBD (80 million tons) would be for Eastern Europe (excluding Rumania and Yugoslavia) and only 1·1 MBD (55 million tons) would be for the Soviet Union itself.[6]

Throughout this period, the CIA position regarding Soviet imports apparently continued to evolve. This was not always evident to the general public, however. Indeed, CIA officials seemed to hold to their pessimistic import predictions and at times ignore even the more moderate July 1977 revision. As late as July 1979, newspaper releases in the *New York Times* and the *Washington Post* appeared to indicate that CIA pessimism was unchanged and that the CIA continued to expect that the Soviet Union and Eastern Europe would be importing major quantities of petroleum by the mid-1980s.[7] Within a week or so, however, the CIA released another report which seemed to back off significantly and conceded that by 1985 there would be no net imports because 'the USSR could not afford to become a net oil importer.'[8]

According to the director of the CIA, this had actually been the CIA's position since late 1977.[9] Much of the confusion is due to the CIA's use of what it refers to as a 'notational gap' approach. This is intended to show what would happen if existing demand and supply trends for petroleum in 1977 were extended mechanically. To complicate the matter even further, the CIA reports do not explicitly state that such an approach is being used; this is something that either has to be surmised or learned from personal discussion. In any case, according to the CIA, the notational gap calculation did not concern itself with what the economic consequences would be of importing that much petroleum. In its defense, the CIA says it just wanted to show what would happen if the trends evident in 1977 continued through into the 1980s. However, this is certainly not how most readers of the CIA report reacted. They assumed that by 1985 the CIA anticipated that annual imports would indeed total the 3·5 to 4·5 MBD figure. What they and even some CIA analysts failed to consider, at least until 1979, was that there was probably no way the Soviet Union and Eastern Europe could find that much spare petroleum in the world market each year. And if the Soviets did find it, there seemed to be no way they could pay for it. What the CIA only belatedly came to realize was that one-half of the Soviet hard-currency earnings come from petroleum exports. If the Soviets had to give up this $6 billion or so in hard currency because they had no petroleum to export, that would leave them with about $6 billion less to buy the 700,000 barrels a day the CIA says they will have to import by 1982.[10] If by 1985, as originally predicted, the Soviet bloc finds it necessary to import 3·5 to 4·5 MBD, then it will have to spend at least $25 billion to $30 billion at 1979 prices just for petroleum imports. In all likelihood this is a considerable understatement since this added demand generated by the USSR and Eastern Europe would lead to a substantial increase in real prices. But even at existing world prices, petroleum imports of $25 billion to $30 billion, when added to their regular imports of about $13 billion, would leave the Soviets with total imports of $38 billion to $43 billion, but with exports of only about $6 billion or $7 billion. The result would be an annual trade deficit of $30 billion to $36 billion. There are not many countries that can handle that kind of trade deficit and there are not many countries that would export to a country like the Soviet Union with that kind of debt burden.

Ultimately, it was calculations of the sort showing that what the CIA had initially predicted was economically impossible that caused the CIA to back off, at least from the import side of its predictions. However, its April 1977 report was more widely publicized than its subsequent reports, especially the 1979 revision, and so not so many, either in the West or in the USSR, were aware of the change in the CIA's position. In any event the Soviets continued to criticize the CIA report and to forecast an increase in production into the foreseeable future.[11] Starting in 1976, they predicted that petroleum production would grow each year until 1980, when it would reach 620 million to 640 million tons. Initially the Soviets made the 640 million ton figure the target, but in the middle of the five-year plan they lowered their sights to the 620 million ton figure. Even if production in 1980 should fall short of the 620 million ton level, which is likely, the Soviets still see themselves exceeding the range of 540 million to 590 million tons which the CIA has predicted.

Are the Soviets correct? Will they be able to avoid the fall in production predicted by the CIA? Will they be able to sustain petroleum exports or will they be forced to import as the CIA at one time seemed to predict? First we shall look at the question of reserves, then production, and finally consider what other options the Soviets and East Europeans have to avoid importing the 2·7 to 4·5 MBD originally predicted by the CIA.

RESERVES

Difficult as it may be to determine petroleum reserves in most regions of the world, it is even more difficult to make estimates for communist countries. Verification of official estimates by outside specialists is normally impossible. In the case of petroleum in the USSR, there are not even official estimates to work with.

According to the Soviet State Secret Act of 1947, the volume of oil reserves is a state secret. To determine the size of Soviet oil reserves, therefore, it is necessary to work with unofficial estimates. These estimates are usually based on an analysis of resource exploitation within the Soviet Union and a projection of the likely oil potential of the sedimentary basins of the Soviet Union measured against the experience of comparable sedimentary basins elsewhere in the world. At best this is an inexact science. Consequently, over the years a variety of unofficial estimates of petroleum reserves have been drawn up. One hazard in this is that several of these calculations stem from the same outside source – the geologist A. A. Meyerhoff. Other estimates vary widely. As a result the observer from afar can only pick a reserve figure that seems most reasonable and logical.

Even if the Soviets released official estimates, there would still be difficulties for the foreign analyst. The Soviets use a classification system which is somewhat different from that used in the West. The Western and Soviet terms correspond approximately (but not exactly) as follows:

Western	*Soviet*	*CIA according to Meyerhoff*
Proved	$A + B$	A + part of B
Probable	$A + B$ + part of C_1	$A + B$
Prospective or		
Possible	$A + B + C_1 + C_2$	
Predicted	$A + B$ + part of C_1 $+$ all of $C_2 + D$	

To make matters more confusing, there is disagreement among Western specialists as to just which Soviet category corresponds to which American category. As interpreted by the CIA, Soviet A reserves are those established through drilling or by fields completely surrounded by wells.[12] B reserves are those surrounded, but not completely, by three producing wells. C_1 reserves are marked by at least two wells in the producing zone or else the area must be directly adjacent to reserves of a higher category. The remaining categories are less defined and therefore less likely to be the site of commercial production. Some authorities, such as the CIA, include part of the B along with all of A in the category 'Proved'. Meyerhoff's definitions are slightly different. He points out that the CIA and others used the outdated pre-1971 classification system, which tends to understate what we, in the West, regard as 'Proved reserves'.[13] According to Meyerhoff, the CIA includes only A category and one-half of B category as Proved reserves. It should have included all A and all of B reserves. Proved and Probable reserves, says Meyerhoff, should refer to A plus B plus part of C_1, which the CIA does not do.[14] Given Meyerhoff's definition, the CIA seems to have understated Soviet reserves. In sum, even the straightforward is complex.

If differences exist over the precise definition of Soviet reserve categories, it is all but inevitable that there will be differences as to the extent of the reserves themselves, particularly given that at least the petroleum reserves are regarded as a state secret. Thus there are a variety of estimates, not all of which specify whether the reserves cover categories $A + B + C_1$ or more.

At the far extreme covering categories $A + B + C_1 + C_2 + D$,

Table 6.1   *Petroleum Reserve Estimates for the Soviet Union, Proven and Probable $(A + B + C_1)$*

Source	Billion Tons	Billion Barrels
1 Congressional Research Service, Library of Congress, CRS-1 IB75059, update, 25 February 1976	5·2–7·0	38–56
2 UN, *Statistical Yearbook 1971*, Vol. 23 (New York, 1972), p. 181	8·1	59
3 CIA, *Research Aid, Soviet Long-Range Energy Forecasts* (Washington, DC, September 1975), p. 10	10·0	73
4 Leslie Dienes, 'Energy self-sufficiency in the Soviet Union', *Current History*, August 1975, p. 47	10·3	75
5 J. H. Cheshire and Miss C. Huggett, 'Primary energy production in the Soviet Union: problems and prospects', *Energy Policy*, September 1975, p. 229	10·5	77
6 *Oil and Gas Journal*, 22 May 1978, p. 33	10·3	75
7 *International Petroleum Encyclopedia* (Tulsa, Oklahoma: The Petroleum Publishing Company, 1975), pp. 203 and 297	11·3	83
8 United States Geological Survey for 1973	13·5	99
9 Jeremy Russell, *Energy as a Factor in Soviet Foreign Policy* (London: Saxon House/ Lexington Books, 1976), p. 40	14·0	103
10 United States Geological Survey, *Summary of Petroleum and Selected Mineral Statistics for 120 Countries, Including Offshore Areas*, Joseph P. Albers *et al.*, (Washington, DC: US Government Printing Office, 1973), pp. 142–3	13·5	99
11 CIA, *Prospects for Soviet Oil Production: A Supplementary Analysis* (Washington, DC, July 1977), p. 32	5·5	40
12 Grant Hopkins, Gulf Oil, 'Oil reserves in the USSR: an assessment', 6 December 1976, p. 13	9·5	70
13 Dienes and Shabad, p. 253	10–11	73–8

Meyerhoff estimates that Soviet onshore and offshore reserves of Proved, Probable and Predicted total 441 billion barrels or 60 billion tons.[15] This is a reduction of his 1972 estimate which was 100 billion tons of petroleum reserves that may some day be recovered. The CIA reports that Soviet geologists estimate total reserves at about 175 billion tons, a figure the CIA considers too high.[16]

Estimates of Proved and Probable reserves are much lower. The Soviets apparently believe they have about 20 billion tons of Proved and Probable $(A + B + C_1)$ reserves of which 10 billion tons are recoverable.[17] Other estimates of Proved and Probable reserves range from 5·2 to 14 billion tons (38 to 103 billion barrels).[18] (See Table 6.1.)

Focusing only on predictions by the CIA yields its own range of estimates. Of course they are gathered at different periods of time, but they diverge by more than the production that has occurred in the interim. Taking just A and a portion of B reserves, the CIA estimated Proved reserves as of 1976 to be about 4·1 to 4·8 billion tons (30 to 35 billion barrels).[19] In 1978 the CIA's estimate of Proved reserves was 35 billion barrels (4·8 billion tons) or at the upper end of this estimate, but this was lowered again in 1979 to 30 billion barrels.[20] For comparative purposes, American Proved reserves total about 30 billion barrels (4·1 billion tons) and Kuwait had about 67 billion barrels (9·1 billion tons).[21]

One thing all specialists, both in and outside the Soviet Union, agree upon is that the discovery rate has fallen and that the Soviet Union is not adding to its new Proved reserves as fast as it should. This is reflected in the officially acknowledged increase in the percentage of C_1 type (Probable) reserves to the total of all reserves. Whereas C_1 reserves amounted to only 33 per cent of the $A + B + C_1$ reserves in 1959, by the mid-1970s they had increased to over 52 per cent.[22]

Those who take a dim view of Soviet production prospects argue that, if the reserves are as low as the pessimists have argued, Soviet production will indeed fall in a short period of time just as US production has fallen, and probably faster.[23] Equally disturbing is the fact that the Soviets have not discovered any new super giant fields recently to compensate them for the fall in output in many of their older fields. Moreover, even if new discoveries are made, it will be some time before the Soviets can start actual production. Thus exploratory geological work becomes all the more urgent if output is to continue to grow yearly and if the reserve-to-output ratio is to be maintained. Since we do not know the size of the reserves, we obviously do not know the size of the ratio, but the Soviets do acknowledge that the reserve-to-output ratio is falling and that they are concerned.[24] Moreover, since it usually takes at least a decade between the time a new field is found and the time output actually starts, the Soviets had better move quickly.

DEPLETION
A major reason for the need for new fields is that the problem of depletion in existing fields is a serious one for the Soviets.[25] Some of their fields have been worked for a half-century or more. In addition, many of the more recently discovered fields were improperly developed. To maximize output in the short run, long-run yields were sacrificed. In theory, that should not happen in a socialist society; rather that was thought to be a shortcoming of the capitalist system. In the conventional wisdom, private property rights precluded the maximization of long-run output. Each property-holder tries to maximize output on his individual holding. That leads to overdevelopment. Consequently there is a loss of underground pressure so that the combined short-run and long-run maximum efficient rates (MER) of output is less than it would otherwise be if a more systematic and co-ordinated drilling and pumping program had been followed. Public ownership of all property as in a socialist state is in principle the best way of co-ordinating that effort and achieving the MER. As on other occasions, Soviet officials became persuaded by their own rhetoric and argued convincingly that yields from their own oil fields would be higher than anyone else's in the world.[26] We saw that one Soviet economist even spoke in terms of a recovery index of 80 to 90 per cent.[27] But while the ideological spirit said yes, the material wherewithal said no.

Despite exhortations by party and industrial officials, Soviet oil workers seemed unable to halt the gradual decline and output in some of their older fields.[28] Petroleum ouptut in the Volga-Ural region, which eclipsed the Baku and north Caucasus region after the Second World War, also began to fall in the mid-1970s. Some of the fields had been producing for twenty-five years. Yet there was not too much concern because as production in the now older Volga-Ural fields declined, production in the newer fields of west Siberia increased at least initially more than enough to compensate for this drop. Whereas production in the Volga-Ural region contributed 59 per cent of total Soviet production as late as 1970, by 1978 its share had fallen to 37 per cent. The residual was more than made up by increased output from west Siberia which in 1970 generated only 9 per cent of total output. By 1978 west Siberian output accounted for about 44 per cent.[29]

WATER INJECTION
While this shift to west Siberia may appear to be good planning, CIA specialists see it as a problem. They assert that the fall in relative and absolute production in traditional areas came sooner than anticipated and was due to poor production technique. According to the CIA, one of the main explanations for the faster-than-anticipated fall in

petroleum output in both the Volga-Ural and west Siberian regions has been the use of water injection in the wells. This is done because, as petroleum is extracted, the gas pressure in the well falls. In an effort to compensate for that fall in pressure, water is injected into the reservoir from surrounding holes in order to increase the pressure underground and push the oil towards the original hole to increase the well's flow. This makes possible a higher output with fewer wells than would otherwise be necessary. It also eliminates the need, at least in the early days, for expensive pumps.[30]

Unfortunately sometimes water is injected at too great a speed and in too large a quantity, especially when the producers are under pressure to increase production. This adversely affects the reservoir. Even when the water is properly injected into the reservoir, the volume of water in the reservoir begins to build up sooner or later and an increasing percentage of water is extracted along with the petroleum. Eventually it is necessary to install submersible pumps to sustain output levels. The Soviets manufacture their own pumps but have found that American pumps are more effective. Since 1971 they have imported more than 1,000 pumps with a lifting capacity of 150 million tons a year to supplement approximately 11,000 of their own pumps.[31] However, because they keep injecting water into the reservoir, more and more pumps are needed as water begins to constitute a larger and larger percentage of the extracted liquid. By the late 1970s, 50 per cent of the liquid lifted in the Volga-Ural region was water. The water cut percentage was also beginning to increase in Samotlor in west Siberia.[32]

The use of water injection extended production considerably in the Volga-Ural region. Even though ultimate recovery from the wells was only a fraction of the anticipated 80–90 per cent hoped for, the yields would have been much worse if it were not for the use of water and the pumps. Under the circumstances, when pressure began to drop at the Samotlor oil fields in the Tiumen region of west Siberia, it seemed logical to introduce the same technique there. But what worked reasonably well in the Volga-Urals worked poorly in Siberia. A. P. Krilov, of the Academy of Sciences, laments that the uncontrolled and excessive use of water injection at Tiumen permanently lowered the area's ultimate output.[33] Thus at Romashkino, the giant producing field in the Volga-Ural region, it took eighteen years before the water content of the fluid being lifted rose to 10 per cent. In contrast at Samotlor with recoverable reserves of about 2 billion tons (Romashkino's counterpart in the Tiumen-west Siberian region), the water cut rose to 10 per cent in only three years.[34] That was much faster than anyone anticipated. To compensate, the Soviets started to increase output beyond the maximum efficiency rate of recovery. Inevitably that only caused the pressure at Samotlor to fall even more rapidly than it would have done with a more prudent pumping schedule.[35] Equally discouraging, while the pumps at Romashkino were good for about a year, the pumps at Samotlor broke down after only two months because of silt and salt.

What happens at Samotlor is critical for the immediate future of Soviet petroleum production. In December 1978 Samotlor was producing at a daily rate of 0·5 million tons or one-third of the overall Soviet output and the increased production at Samotlor was being used to compensate for the fall in petroleum production at the older fields in the USSR.[36] But given the increasing water cut at Samotlor, the CIA questions how much longer production there can increase, particularly if water injection continues to be the most important method of boosting or enhancing production. The CIA calculation is a simple one. The Soviets claimed that in the beginning Samotlor had a total reserve capacity of 2 billion tons. By April 1978 the Soviets reported that 500 million tons had already been removed.[37] Therefore, if the Soviets keep withdrawing 150 million tons a year as they were doing in the late 1970s, that will leave only ten more years of productive life at best. But as the reserve level drops, so does the pressure and so should output. As a result, output for the whole Soviet Union will drop by the almost 30 per cent mentioned earlier in this chapter, or so the CIA and some Soviet authorities contend.

FINDING NEW FIELDS

Not only have the Soviets had a hard time maintaining production in existing wells, they are also having a hard time finding new wells. Here, too, ideology has misled the Soviets. They have come to believe that in the capitalist world oil field producers overemphasize short-run profits at the expense of long-run exploration. This is not entirely correct. Part of the Soviet misconception is due to the fact that the Soviets are relatively unfamiliar with the capitalist concept of discounted present value. Since the Soviets do not work with discounted income streams, they generally do not appreciate how it induces private oil producers to spread their earnings over a long period of time.[38] (It should be noted, however, that even with the use of future discounted value and the effect that it has on prolonging the life of a producing field, there are some in the West who would argue that such a system still lessens the productive life of oil wells beyond what is possible from an engineering point of view. As a minimum, unless all the mining rights to a particular deposit are controlled by a single company, there must be a regulatory commission or some co-ordinating body to insure that one owner does not try to draw out his neighbors' share of the oil.)

Freed from the need to worry about profit-making and fluctuations in prices, Soviet petroleum engineers, in contrast to their Western counterparts, supposedly should be able to concentrate solely on maximizing oil production over the longest period of time. What is forgotten, however, is that while Soviet petroleum officials do not have to enslave themselves to the short-run task of increasing the bottom profit line, they still have another master who tends to be just as demanding and distorting: the need to fulfil the highest possible five-year plan output target.[39] The emphasis continually is on increasing output in the existing five-year plan period even though output can only be maximized if it is spread instead over thirteen or even seventeen years. That means that the oil drillers will tend to focus on increasing petroleum reserves in the short-run period for which premiums can be awarded and enjoyed now rather than in some undefined future. This became a particularly attractive option once it appeared that the days of finding massive new oil fields had passed. Thus, whereas in 1964 and 1965 Soviet oil output increased by 13·6 million tons for every million meters drilled, in 1976 the figure had fallen to 5·6 million tons.[40] In part this is because the drillers tend to spend most of their time on drilling development wells, that is, wells in already producing fields as opposed to exploratory wells where there is no already producing well nearby. For example, the number of meters drilled in exploratory wells actually fell from 5,802,000 meters in 1967 to 5,418,000 meters in 1975.[41] Similarly in the Tiumen region, 25–30 per cent of capital investment spent on geological prospecting goes to exploratory drilling. While this may seem adequate, if major new deposits are to be found, it should be 60–70 per cent higher.[42] What the Soviets really need is the discovery of another giant oil field similar to Romashkino or Samotlor. Instead, because of the inattention paid to exploratory drilling, they are only finding smaller oil fields in more remote locations scattered around.[43] Nor can they continue this way. By 1978, the fields of Samotlor, Mamontov, and Ust Balyk had all reached their rated capacity. This means that they cannot be counted upon to provide ever-increasing offsets to production depletion elsewhere.[44]

The shortcomings in the drilling program are not due solely to short-sightedness on the part of Soviet geologists. Much of the difficulty is due to the MPI's inability to supply enough manpower and high-quality equipment to the remote areas of the north and Siberia, which are plagued by mosquitoes and swamps in summer and bitter cold in winter. Moreover, many of the supply routes and facilities must be built in permafrost. It is hard to attract large numbers of workers to these areas.[45] This in turn is partly due to the difficulty of building up an infrastructure to house and supply them. The Soviet press is filled with one story after another about the failure to fulfill supply and construction plans on time.[46] When highways, railroads, and pipelines are not built according to plan, the goods must move by barge or plane.[47] One takes too long and is a seasonal affair and the other is very costly and sometimes hazardous.

Soviet petroleum officials have concluded that the supply problems are so serious that they should change their strategy. Instead of attempting to build large cities under such difficult climatic and geographical circumstances, they fly in shifts of workers from base camps in more accessible regions.[48] These workers are flown in for work-periods of one to two weeks, housed in dormitories, and then flown home again. This system is not unique in the Soviet Union. Canadian corporations have also adopted similar systems for much the same reasons.[49] However, while it is claimed this system is very efficient and lowers cost, there are difficulties in maintaining continuity when the work-crews turn over completely every few weeks.[50]

Periodically the government focuses national attention on the difficulties facing the energy industry by calling for increased production and decreased consumption. Sometimes such campaigns are part of a general effort to conserve raw material usage, as in 1972 and 1975, and in other cases the campaign focuses specifically on energy, as in 1974, 1976, and 1977.[51] After the record cold wave in the winter of 1978–9, when natural gas pipelines cracked and trains stopped running, demand for fuel oil often exceeded local supplies. Not surprisingly the Central Committee of the Communist Party issued a decree on 14 June 1979, calling for extra effort to prevent a recurrence of the energy shortages during the 1979–80 winter.[52] That such decrees are ineffective is suggested by the fact that it is necessary to keep passing the same decrees.

Undoubtedly there is much for Soviet petroleum officials to concern themselves with. These problems are also reflected in the CIA reports. They are real, and in the context of the existing situation they will not be easily solved. However that is not the same as saying that there is no solution. It should be recognized, however, that solutions may require not only ingenuity and large sums of money, but also a radical change in Soviet methods, a new way of operating.

ENHANCED RECOVERY

Presumably a complete overhaul of the whole planning system would go a long way towards postponing the depletion of Soviet wells and increasing the discovery of new wells. However, there is another important but less radical approach. The Soviets can use other forms of enhanced recovery technology, and if they cannot invent this technology themselves they can buy it.

Only recently have the Soviets used other methods besides water injection, the system which plays a big part in the CIA's pessimistic scenario. Although in 1977 water injection accounted for 86 per cent of such secondary recovery procedures, at best it provides a yield of 40 per cent and then probably not very often. An alternative method that the Soviets have used is steam injection. This is particularly useful when breaking up high-viscosity crude oil where water injection is particularly ineffective.[53] This same process is also used successfully in the United States. Other enhanced recovery techniques include the application of detergents and carbon dioxide and the injection of natural gas into the reservoir instead of water. The nature of the petroleum reserve in each reservoir determines which process is used, but skilled application of the proper technique can raise yields to 50 per cent. Some Soviet specialists claim secondary recovery can increase annual output by 60 million tons a year.[54]

Hitherto, a major obstacle to the Soviet use of such procedures has been that Soviet industrial ministries and machinery manufacturers were slow to produce the necessary equipment.[55] The deputy minister of the petroleum industry, E. Khalimov, charges that the Ministry of the Chemical Machinery Construction Industry and the Ministry of Energy Machinery Construction produce only 2 per cent of the equipment needed. Though the Soviets claim that they were among the pioneers in the development of the gas lift process which injects gas instead of water to supplement reservoir pressure, Soviet enterprises do not produce the equipment necessary to use the procedure extensively.[56]

While attention is focused on the difficulties the Soviets have had in moving beyond the water injection system to more effective methods, critics like the CIA sometimes disregard the fact that, once adopted, some of these methods have enormous potential. More important, the Soviets are gradually, although belatedly, moving to utilize such techniques, even if it means having to import equipment costing hundreds of millions of dollars from the United States and other capitalist countries. Thus output in some areas such as Grozny and Baku, after falling for years, increased again once gas lift equipment was installed. The same thing happened in Sakhalin.[57] The purchase of a secondary system from one American company for use in the Caucasus has increased yearly output of some of the older Baku wells threefold, from 300,000 to 1 million tons a year.[58]

Recognizing that water injection is not suitable for the Samotlor and Fyodorovsk regions, the Soviets have signed a massive $226 million contract with a French company to install a gas lift system. It will be worth the expenditure if it increases the overall long-run output of the reservoirs in the area and helps the Soviets break out of the trap of having to rely on water injection. Recognizing this, the Soviets have moved rapidly to utilize this technology. Only a year after the contract had been signed, they claimed that the new enhanced recovery methods were already being used in Samotlor.[59] In other areas, however, where water injection still seems to be the best solution, they are simply importing more submersible pumps.[60]

From all indications, the Soviets began to devote an impressive amount of attention to the problem of enhanced recovery sometime in late 1977. Not only did they begin discussions about importing gas lift equipment, they began importing a wide variety of other technology as well. Thus in late 1977 they signed a $23 billion contract with Pressindustria of Milan, Italy, for a plant to manufacture 250,000 tons of non-ionic surfactants a year for enhanced recovery.[61] A few months later they spent $36 million to buy two carbon dioxide liquefaction plants from Borsig, a West German subsidiary of Deutsche Babcock. Struther Wells sold a similar plant for $25 million.[62] Carbon dioxide, like natural gas, serves to increase the pressure in the reservoir. By 1978, petroleum equipment imports had reached a new high.[63]

Because the Soviets lag behind much of the rest of the world in the use of enhanced recovery techniques, that does not necessarily mean that the rest of the world, particularly the United States, has fully utilized enhanced recovery methods. There has been criticism in the United States about how slowly we, too, have moved. The major reason for this was that for some time American governmental price controls made it uneconomical. Without price decontrol, or at least a substantial increase in prices, it was not worth the extra cost of buying the detergents or installing the extra equipment.[64] Ironically, some Soviet specialists have put forward the same argument. As they see it, the existence of different prices for different producing areas in the Soviet Union discourages the use of enhanced recovery techniques, particularly when producers in one area receive a lower price than producers in a neighboring area.[65] They would like a unified sale price system. Without acknowledging or perhaps even being aware of what they are seeking, these Soviet specialists seem to be asking for a system equivalent to one that would utilize marginal cost pricing.

While these various shortcomings indicate how wasteful the Soviet production system is, they also indicate something else: the Soviets probably have immense potential for enhanced recovery. The CIA is skeptical. It reports that in 1976–8 only about 0·4 per cent of total Soviet production was due to enhanced recovery.[66] The figure had increased to 2 per cent in 1979 but there are significant problems ahead if the Soviets are to increase the total significantly. Each field has its own peculiarities and, therefore, must be custom-fitted with its own special enhanced recovery technology. This takes time, patience, skill, and enormous sums of money. Certainly the Soviets have a long way to go before they have the incentives, know-how, or the equipment to take full advantage of the potential that is available. Nevertheless, they have started to respond. Failing to recognize this potential, as the CIA sometimes has, leads to an underestimation of Soviet capabilities.

MORE EFFECTIVE DRILLING

There seems to be a similar lack of appreciation of the potential the Soviets have for increasing their discovery of new fields. True, before there can be much progress there must be substantial improvement in Soviet drilling procedures. While the Soviets cannot revolutionize their traditional methods overnight, the fact is that just as they are increasing their use of enhanced recovery procedures, the Soviets are also taking steps which could increase the extent of their exploration work significantly. One of their biggest problems is that they have limited drilling capabilities. As we saw, they take a long time to drill, they cannot drill large numbers of deep holes, and they have an incentive and planning system which often leads them to drill the wrong depth hole in the wrong place. But the Soviets have finally come to recognize their failings.[67] Of course recognition in and of itself is not enough, but it is a crucial first step towards remedying their problems.

Perhaps the hardest of all challenges is to streamline the incentive and planning system. Since this brings them up against the very essence of the Soviet system, the chances for an overall reorganization are slight. Nevertheless, there have already been some modifications. The Soviets, for example, have recognized that with the exception of the shallow areas of the Caspian Sea, they have not been able to utilize their potentially rich offshore areas. Their approach to this problem typifies the way in which they tend to respond.

First, they have realized that the division of responsibility among so many agencies made it easy to avoid responsibility. To provide for better co-ordination and accountability they have sharpened their control and lessened some bureaucratic conflicts. For example, traditionally offshore work was divided among the Ministries of Geology, Gas, and Petroleum, as well as an organization called Gostekh nadzor, the State Mining and Technology Control Agency.[68] Each one went about its own business without much in the way of co-ordination. Little was accomplished beyond shallow drilling in the Caspian Sea and that dates back to 1929.[69] Although it can hardly be regarded as a solution, from January 1979, the Council of Ministers concentrated all responsibility for offshore exploration and drilling in the hands of the Ministry of the Gas Industry.[70] It will be surprising if drilling for petroleum gets as much attention as drilling for offshore gas, but at least the concentration of bureaucratic responsibility should improve co-ordination and reduce jurisdictional overlap. There is also likely to be a similar reordering of deep drilling responsibility, another area where the

USSR has lagged badly.[71] Most important, the Soviets have also begun to change the incentive system in order to induce more rational and effective prospecting and drilling procedures.[72]

In addition to reorganization, the Soviets have also availed themselves of better technology. As we have seen, when domestic industry cannot do the job, the Soviets often turn to imports. Thus they have bought semi-submersible offshore oil rigs, and the technical know-how to build such structures themselves. These rigs will operate not only in the deeper parts of the Caspian and Black Seas, but in the Barents, White, and Kara Seas of the north. Until now the Soviets have held back development of these resources because they lacked adequate technology. Hitherto, because of their traditional reluctance to involve foreigners in such projects, they did not seek foreign help. Now, because of the seriousness of their need for increased production, they have turned to foreign technicians and foreign technology.[73]

The Soviets are not worried about lack of oil fields to explore. In fact, it was not expected that the present giant fields in the Mesozoic zone of west Siberia would become the main producing area so soon.[74] Until the 1950s the Soviet specialists believed that the deeper Palaeozoic strata would prove to be more productive than the Mesozoic. But once the big discoveries were made in the Mesozoic zones of west Siberia, work in the Palaeozoic formation was downgraded, as was exploratory work in east Siberia and the northern regions.

While much of the Soviet Union has already been explored, most of that exploration is relatively primitive largely because Soviet geologists lack some of the more advanced seismic technology recently developed. After all, the Soviet Union contains 37 per cent of all the earth's sedimentary areas, including the continental shelves, supposedly the areas of greatest potential. Western geologists point out that the West Siberian Basin is 'one of the largest sedimentary basins in the world.'[75] For comparison's sake, the Middle East has 11 per cent and North America 2 per cent of such sedimentary areas.[76]

Again the Soviets hope to pursue their exploratory efforts primarily with better foreign technology. Their own equipment is either in short supply or inadequate. Soviet exploration equipment is on a par with equipment used in the United States in the 1950s and operates poorly in depths greater then 2,000 meters. Consequently in the late 1970s they purchased several units for mapping geophysical information from such American companies as Geosource.[77]

Most frustrating for Soviet petroleum exploration has been the lack of high-quality drill bits and better quality pipe. We saw that Soviet exploration teams take so long to drill partly because of pipe breakdowns and partly because the drill bits are of such poor quality that they must be replaced continually. These handicaps limit the drilling range of most Soviet efforts to under 2,000 meters.[78] That was adequate in west Siberia where oil was found at a depth of 1·6 to 2·4 kilometers.[79] But in future fields, such as the deep part of the Caspian Sea, the Black Sea, Timan-Pechora and the Viliui Basin where the greatest promise seems to be at depths of 2,500 to 5,000 meters, the Soviet turbo-drill and seismographic system cannot penetrate.[80] At the present time, almost no meaningful drilling takes place beyond 3,000 to 5,000 meters. To the half-empty pessimist, this indicates overwhelming difficulties; to the half-full optimist, it suggests Soviet potential, particularly now that the Soviets are buying foreign seismographic equipment, foreign drill bits, and the advanced American drill bit plant from Dresser Industries. The new drill bits will make it possible to drill deeper and faster with the existing turbo-drill system. The Soviets have also purchased foreign steel mills capable of producing higher quality steel, some of which will be used to make better drilling pipe.[81] The next step may well be the purchase of a foreign rotary drilling plant.

While it will be some time before the Soviets are able to put all their new products and techniques to work, there have already been some promising beginnings. Moreover, they are using their new purchases in efforts to increase considerably the amount of exploratory and development drilling. In west Siberia the amount of drilling was projected to double from 5 million meters in 1979 to 10 million meters in 1980.[82] Without adequate supplies and infrastructure, it will be difficult to fulfill the plan. Still, the volume of exploratory drilling has already started to increase.[83]

As yet nothing comparable to a second Samotlor has been found. It may never be found, but there are hopeful signs. Until now most of the Tiumen output of west Siberia has been drawn from the Mesozoic zone under the most difficult circumstances. New discoveries include commercial finds in the deeper Palaeozoic strata of west Siberia.[84] The Palaeozoic finds, although deeper, are more conveniently located to population centers and are less likely to be affected by swampy or permafrost conditions. Increasing drilling depth should also make it possible to drill more readily under existing gas deposits to new petroleum depostis. Already new deposits have been discovered under the north Tiumen gas fields at three different locations.[85] Until 1979, no petroleum in Tiumen had come from depths greater than 4,000 meters, but there is reason to believe that this deeper region will be the site of vast new discoveries, many of which will not be too far from the existing pipeline infrastructure.[86] Other areas with promise include the Baltic Sea, East Siberia, Timan-Pechora, Tomsk, Yakutia, Irkutsk, Krasnoiarsk in the Ukraine, and Surgut, which is expected to be the workhorse of the future.[87]

As a sign of how serious the Soviets are about developing some of these more difficult sites, they have even authorized joint development ventures with foreign concerns on Soviet territory. Soiuznefteexport's many deals demonstrate the Soviets are not, in principle, opposed to joint ventures. But since the Second World War, these operations have always been outside the Soviet Union. By 1945 all the foreign firms which had been allowed to operate under the Soviet Union had been sent home. While the Soviets have allowed one or two foreign companies to operate in the Soviet Union since then, most of these arrangements have been for a short time without any provision for sharing profits or royalties. There has been talk of an ongoing production and royalty relationship with the Bendix Corporation, but so far nothing concrete has come of that. Thus, making an exception for petroleum exploration suggests that the search for petroleum has very high priority.

Until recently the most ambitious joint venture has been off the coast of Sakhalin. There in 1977, after a three-year search, the Japanese Petroleum Development Corporation finally found petroleum and gas.[88] This is a Japanese-led consortium with a 7 per cent interest held by Gulf Oil.[89] The Soviet contribution is the worksite and labor. As a measure of its confidence that it would find petroleum and that the site will prove commercially profitable, the consortium has borrowed over $170 million to finance its activities. Assuming that the newly found petroleum is ultimately saleable, the foreign group will split the revenues on a fifty-fifty basis with the Soviets.[90]

Another venture involves a Canadian–Soviet effort in the far north autonomous republic of Komi. Called Vosei 100, this group has been drilling wells since 1977.[91] The primary purpose of this venture is to test Canadian drilling equipment in permafrost conditions. Negotiations are also under way with some other foreign companies, including British Petroleum, Phillips Petroleum and Armco. In a manner not seen since the 1920s, the Soviets are negotiating contracts with these companies to set up operations on Soviet territory in the deeper parts of the Caspian and Barents Seas. Reportedly a British and a French company are already drilling in the Barents Sea. Onshore operations may include the Baku region and even the Volga-Ural region. The Soviets wanted to sign a contract to this effect in 1979 but because of the technicalities had to wait until 1980. The expectation is that some of this work, particularly that done offshore, can produce immediate results in as little as two years. Such a time-sequence is not unrealistic since even less time was needed when similar operations were undertaken in the North Sea by some of these same companies.

Some of these projects will involve the sharing of any ultimate production, particularly if the finds are of petroleum as opposed to natural gas. Given Soviet sensitivity to foreigners working on their territory and their past rejection of profit-sharing joint ventures, the fact that they are so intent on making an exception in this instance suggests again the importance to the Soviets of discovering and exploiting new sources of petroleum. If necessary, they are prepared to compromise some of their ideological principles to facilitate that search.

However, even if some of these efforts lead to the discovery of petroleum, this will not necessarily solve all the Soviet problems. The size of the country increases the odds that petroleum will be found, but it also means that the deposits may be too remote from existing pipelines and consumers to make them commercially feasible. That is one reason why enhanced recovery is such an attractive option. Since the existing wells, pipelines, refineries, and markets are already linked up, no new major capital investment is necessary.

Those who predict bleak prospects for Soviet energy often point out that as the Soviets move farther north and east in search of petroleum and other energy sources, the costs skyrocket making the development of these new deposits uneconomical. They suggest that these costs may negate the usefulness of some of the new petroleum finds.[92] Indeed, the cost of oil production from 1959 to 1978 rose by 46 per cent or more and gas by 100 per cent.[93] Costs reportedly doubled from 1970 to 1980. A large part of this increase is due to the rapidly increasingly size of the capital investment needed. Some estimates indicate investment costs may have increased by 50 per cent from 1970 to 1980.[94] Considering the remoteness of the new fields and the difficulty of working in such unpleasant and harsh conditions, the higher costs were to be expected. Given the greater distances involved, it is no surprise that transport costs and thus the delivered energy costs have gone up.[95] Also, to the extent that expenditures have been necessary for enhanced recovery, the profitability of many of the older fields has also fallen.[96]

As the older sites in the European part of the Soviet Union were depleted, and more and more of the country's raw materials had to come from the north and east, raw material costs to the users in the western part of the Soviet Union have increased. Nor is the situation helped by the fact that 75 per cent of the Soviet population and 83 per cent of its industrial population is located west of the Urals, whereas only 10 per cent of the mineral fuel reserves are there to service that population.[97] The location of most reserves has always been inconvenient, but until recently there were reasonably abundant reserves west of the Urals. Thus even as late as 1975, the European sector of the Soviet Union provided 50 per cent of the country's fuel and energy.[98] However by 1980, as the western reserves were depleted, the European portion of the Soviet Union could supply only 37 per cent and the Urals 7 per cent, while the eastern part supplied 56 per cent of

Soviet energy. Based on the present knowledge of existing reserves, there is every reason to believe this eastward trend will continue.

Yet while Soviet planners might prefer more readily accessible supplies, the remoteness of the new supplies does not necessarily mean that raw material production is about to become unprofitable. One way to cope with the problem is to move the consumers of energy to where the energy resources are. Of course, this means bringing workers, housing, and infrastructure with them, which is expensive and which the Soviet population does not find particularly appealing. Yet, since the expected cost of energy resources such as the open pit coal or lignite at Kansk Achinsk and Ekibastuz in Siberia are calculated to be about 35 per cent of the cost of an equivalent amount of Tiumen oil or gas, and 15 per cent of the cost of the deep pit Donetz coal, there is a considerable margin to work with.[99] (See Table 6.2.) It should be pointed out, however, that while the price advantage of using Siberian coal is enormous, the coal at Kansk Achinsk, while available in large quantities, is heavy in ash content (low in heating value) and easily self-ignited. Finally, Soviet pricing procedures cause confusion, if not mismanagement. Because many capital and geological costs are not fully included, there is no assurance that these figures are comparable to Western figures. This may mean that Soviet energy resources may be more expensive to produce than Soviet figures suggest.

Table 6.2   *USSR Estimated Cost of Production of Soviet Fuels in 1975*

Site	Rubles per ton of standard coal equivalent 1975
Coal, open pit, Kansk Achinsk	2·4
Coal, open pit, Ekibastuz	2·5
Natural gas, Tiumen oblast	6·6
Natural gas, Central Asia	7·0
Crude oil, Tiumen oblast	6·7
Coal, open pit, Kuznetsk	8·6
Coal, underground, Karaganda	12·8

Source:
Mazover, 1975, p. 66.

But whether the figures are valid or invalid, the Soviets are moving an increasing percentage of their manufacturing facilities to Siberia. By their calculations, even with coal that is cheaply mined, the cost of transportation is so high that it is economically irrational to ship coal from Kansk Achinsk in Siberia to the central regions of the Soviet Union.[100] Some of their proposals for the development of Siberian energy on site are quite dramatic. Although there are serious technical obstacles to master, one of the projects calls for the building of large on-site power generating plants at both Kansk Achinsk and Ekibastuz. The Soviets expect to construct massive strip mines from which the coal and lignite would be taken to thermal power stations sited close to the pit mouth.[101] Like the Four Corner generating plant in the United States, the electricity would then be sent on to more urban locations. The Soviets hope to use high-voltage transmission lines, some with a capacity of 1,500 kilovolts. Since the energy on site is so cheap and readily available, this makes sense. Given the pollution problems the transportation and burning of such low-quality lignite coal would generate, it clearly is better to burn this material at the mining site in Siberia than in more populated areas.[102] However there are still some engineering challenges to overcome. Because of its complex nature, Soviet engineers must learn how to control the burning of the Siberian lignite coal. They will also have to work out some of the problems associated with the transmission of electricity over such long distances. The Soviets have made great strides along these lines, but much remains to be done.

The Soviets are also investing enormous sums of capital to build a series of Siberian petrochemical complexes. The largest effort is centered in Tomsk. It will utilize the region's oil and gas and produce a vast array of petrochemicals, including polypropylene, methanol, urea, formaldehyde, polyformaldehyde, ethylene, and polyethylene. A large number of these plants will be built and equipped by Western manufacturers.[103] Although there are many construction difficulties, the Soviets anticipate that the region will become one of the largest petrochemical complexes in the world.[104]

While it may be economically sensible for the Soviets to move their industry to Siberia and utilize their energy raw materials there, this relocation of industry does not address the question of whether or not the rising costs of Soviet energy will make it unprofitable for the Soviets to export petroleum. Again, any attempt to make a judgement on this matter presumes that Soviet costing and pricing procedures are the same as ours. Generally, as we saw, they are not. First, we saw how not all geological survey costs are covered. Although the Soviets recognize the problem, and are now covering more of the costs than previously, the government must cover the costs in some other way, including occasional subsidies.[105] Even more important, Soviet wholesale prices in 1979 did not fully reflect the increasing capital costs necessary to work the country's increasingly remote mining sites.[106] Of course a major oil strike lowers the unit cost of any remote oil discovery, but there have not been any for some time. Consequently the profitability of the Soviet petroleum industry continues to fall.[107]

While mindful of the shortcomings of the data, the Soviets have published enough information so that we can make a reasonable calculation of the profitability of Soviet petroleum exports. According to the Soviet economist A. Probst, in 1971 the cost of sending petroleum equivalent to 1 ton of conventional fuel 2,000 kilometers by pipeline, was 88 kopecks.[108] Since the distance from the west Siberian fields of Tiumen is almost 3,000 kilometers to the Black Sea, or 4,000 kilometers to the Baltic, and since 1 ton of crude oil equals 1·43 tons of standard fuel equivalent, this means it should cost no more than 1·9 to 2·5 rubles to send a ton of oil by pipe to the Black Sea on the western Soviet border. Since the 1975 cost of a ton of petroleum measured in tons of standard fuel equivalent was 6·7 rubles, up slightly from the 1971 figure of 6·2 rubles reported by Probst, the 1975 cost to the Soviet Union of a ton of west Siberian petroleum transported to the Soviet border approximates 11·5 to 12·5 rubles a ton.[109] That, in turn, is equivalent to 1·6 to 1·7 rubles (about $2·10 and $2·30) a barrel at 1975 prices. Of course in the early 1970s west Siberian petroleum was not as important as it was to become later in the decade. The Volga-Ural region, a shorter distance to the border, was the main producing area. Still this is a reasonable approximation of costs in both 1971 and 1975 because production of a ton of petroleum in west Siberia was so cheap. In any case, given that the Soviet export price to Italy in 1971 was about 13 rubles a ton, there is some reason to wonder how much, if any, profit the Soviet Union made on such transactions. This also explains why the Soviets were so insistent on charging more to most of their East European customers.

Of course, determining how much the Soviets actually charge for their exports is very complex. Since the Soviets do not allow for free convertibility of the ruble, the rate of exchange set by the Soviets between the ruble and other currencies is an arbitrary matter. Moreover, it is widely accepted that the ruble is overvalued. The official rate of exchange with Western currencies is arbitrarily determined by the Soviet government. Since virtually all Soviet export transactions are first recorded in dollars or some other hard foreign currency, the price of an export, which is then translated into inconvertible and overvalued rubles, may bear little relationship to reality. Thus using the export ruble figure often leads to all kinds of confusion. For example, even though it cost the Soviets in 1971 about 11·5 to 12·5 rubles to produce and ship a ton of petroleum which was then sold at only about 13 rubles, it may not have been a foolish transaction. The fact that the sale of this petroleum made it possible for the Soviets to import Western machinery and pipe which otherwise they would probably have been unable to buy casts a different light on the transaction. If the Soviets had been unable to export petroleum, the likelihood is that they would have had to sacrifice considerably more in domestic rubles to find Soviet goods acceptable as payment to the Italians. For that matter the Soviets may have had no substitute. Thus the export of 1 ton of petroleum for 13 rubles may have made it possible to import Italian goods that would have otherwise cost the Soviets 30 to 45 rubles to produce at home. In these circumstances, it paid the Soviets to export petroleum for hard currency even though at the official rate of exchange it looked as though their ruble import earnings amounted to less than the cost of producing and shipping their crude oil for export. If forced to, the Soviets might even have been prepared to reduce their export prices to the point where their hard-currency earnings on petroleum just barely exceeded the hard-currency expenditures necessary to prepare that petroleum for export. Since a large percentage of the pipe used in the Soviet Union is imported, this may be a significant sum. However the oil itself and the labor are all paid for in soft ruble currency and therfore the ruble costs incurred do not necessarily have to be covered by foreign exchange earnings.

While hard-currency exchange earnings do not have to cover all the ruble or soft-currency expenditures, needless to say the Soviets would like to cover them. Prior to 1973, there was some concern that the Soviet Union might have been losing money on its petroleum exports.[110] That is, it may not even have been recouping that part of the cost that was explicitly included as part of its costs. This was probably particularly true about exports to Eastern Europe where the ruble rate of exchange was not so overvalued. Thus the official ruble earnings of petroleum exports should have been high enough to cover domestic ruble costs. That this sometimes was not the case had been candidly conceded by the Soviet economist Iu. Iakovets, who wrote that 'Until recently the export of several types of raw materials was unprofitable. With the increase in world prices on raw materials, the export effectiveness of these goods increased.'[111]

The 1973 embargo and fourfold price increase changed the situation. There can be little debate that by the time west Siberian petroleum or other types of Soviet petroleum made their way to the Mediterranean or Baltic Seas, it was not cheap oil, as least as measured by the cost of pumping and shipping Middle Eastern oil. However, after 1973 it did not matter so much. The higher price after October 1973 demonstrated what economists had been laboriously arguing in their classrooms all these years: raising prices makes profitable what previously was unprofitable or marginally profitable. If that holds for the North Sea and Alaskan oil, which was generally unprofitable prior to 1973, it holds for west Siberian oil.[112] Based on the reports of the Soviet cost of production and transportation in 1971, the price increases set in motion in 1973 provided the Soviet Union with a very comfortable return on its exports to the hard-currency world. From that point on Soiuznefteexport, the Soviet agent for the Ministry of Petroleum Industry, was assured of a very handsome profit on the exports from virtually all working Soviet oil fields. Such high prices also provided a most comfortable margin for future exploration even though the fields may have been more remote than those in Tiumen.

However, what was good for Soiuznefteexport could not always be fully enjoyed by its principal, the Ministry of the Petroleum Industry. The world price of petroleum may have quadrupled (at least to the hard-currency countries) but internal prices and prices to CMEA remained the same. Even when CMEA began to pay higher prices in 1975, according to Soviet data the internal wholesale price in the Soviet Union for fuels and in all probability petroleum actually seems to have fallen.[113] This resulted in a steadily diminishing flow of profits both in rubles and as a percentage of return.[114] However, this would not have been the case if the Ministry of Petroleum Industry had been permitted to reflect the foreign exchange earnings resulting from its endeavors. Some Soviet economists have come to recognize what is happening. While they have been reluctant to say outright that faulty price policy and the segregation of hard currency and CMEA petroleum export revenues from the revenues derived from the domestic market causes a misallocation of resources, they do acknowledge that export earnings may sometimes not receive enough recognition. Furthermore, they also seem to have come to the realization that because of the unduly low price on domestic petroleum, consumption of petroleum in the USSR may be higher than it should be given the opportunity cost of the lost export earnings.[115] For that reason, Soviet authorities were expected to raise all wholesale prices in the early 1980s, the first such revision since 1967.

CONCLUSION

Exactly how large workable Soviet petroleum reserves are is unclear. The CIA, for example, uses a classification system that tends to understate what the Soviets have. Furthermore the Soviets potentially have vast reserves that are as yet not surveyed. Given the proper incentives and tools, the Soviets stand a good chance of discovering new fields and new areas as well as deeper fields in existing areas. Considering the difficult nature of the Soviet terrain and climate, and the almost counter-productive nature of the Soviet incentive system, what is surprising is not that they have not discovered more, but that they have discovered as much as they have. Moreover, they have discovered and brought these fields to fruition in a remarkably short period of time, Western skepticism notwithstanding.[116] What we sometimes forget is that the Soviets often work under circumstances that many Americans would find intolerable. They may be less productive, and there may be enormous waste associated with their efforts, but they can marshal vast resources and large numbers of slow but patiently plodding workers when they want to.

If they can obtain the foreign technology, the Soviets should also be able to avail themselves of some significant output gains through enhanced recovery techniques. While discoveries of new fields will ensure increased output a decade or so from now, enhanced recovery should make possible a more immediate increase in output. It is true, as the CIA says, that in many cases water injection methods have been overly and poorly utilized. However, just as foreign producers have found other ways of increasing their output, so the Soviets are now seeking foreign help in using these methods within the Soviet Union. Enhanced recovery will not prevent the eventual depletion of a field, but it will extend its productive life, at least in the short run.

If petroleum prices had not taken the great leap upward that they did in 1973, most of the skeptics would probably have been proven correct. But at current world prices, the Soviets can afford to push even farther out into the Siberian wastes and waters and at the same time import the foreign technology needed to enhance output. Thus the Soviets probably have the wherewithal to sustain petroleum production and conceivably increase it. Exports will fall at some point, but there is good reason to suspect that it will not be quite as soon as the CIA says, and even if exports do fall, the periodic increases in world prices make it possible for the Soviets to earn the same amount of foreign exchange while exporting less.

But what if output levels off or even falls? Will the Soviets and their allies find it necessary to import petroleum in the quantitites indicated by the CIA? That would gut their export earnings and force a slashing of their imports. This in turn would be bound to have a devastating effect on their economy, if for no other reason than that they would be unable to pay for their grain imports. In the circumstances, even though output might fall, the Soviets can be expected to exert enormous effort to divert petroleum from domestic needs in order to sustain their exports. Is that possible?

NOTES: CHAPTER 6

1 CIA Soviet, April 1977, pp. 1, 2 and 9; CIA, *The World Oil Market in the Years Ahead*, ER 79-10327U (Washington, DC, August 1979), p. 37.
2 *New York Times*, 30 July 1979, p. D1.
3 CIA, *Simulations of Soviet Growth Options to 1985*, ER 79-10131 (Washington, DC, March 1979), p. 5.

4 CIA, March 1979, p. 2; CIA Soviet, April 1977, p. 1; CIA, August 1979, p. 37.

5 CIA, *Soviet Economic Problems and Prospects*, ER 77-10436U (Washington, DC, July 1977), p. 22 (hereafter CIA Economic, July 1977).

6 CIA, March 1979, p. 6, Table 2; CIA Economic, July 1977, p. 22.

7 *New York Times*, 30 July 1979, p. D1; *Washington Post*, 30 July 1979, p. A4.

8 CIA, August 1979, p. 40.

9 Personal letter from the director of the CIA.

10 *New York Times*, 30 July 1979, p. D1.

11 *Sovetskaia Rossiia*, 31 August 1979, p. 3.

12 CIA, July 1977, p. 31; John Hardt, Ronda A. Bresnick and David Levine, 'Soviet oil and gas in the global perspective', 'Oil and natural gas reserves of the Soviet Union and methods of their estimation', in *Project Interdependence: US and World Energy Outlook Through 1990* (Washington, DC: Congressional Research Service, Library of Congress, US Government Printing Office, November 1977), ch. XXIX, Appendix 2, p. 827.

13 *Review of Sino-Soviet Oil*, December 1977, p. 12.

14 ibid., April 1979, p. 19a; Russell, p. 40.

15 *Review of Sino-Soviet Oil*, April 1979, p. 19a.

16 CIA, *Research Aid, Soviet Long-Range Energy Forecasts*, A(ER) 75–71 (Washington, DC, September 1975), p. 10.

17 ibid., p. 10.

18 Hardt, Bresnick and Levine, p. 844.

19 CIA Oil, July 1977, pp. 32–3.

20 CIA Handbook 1978, p. 84; CIA, August 1979, p. 38.

21 CIA Handbook 1978, p. 84.

22 Leslie Dienes and Theodore Shabad, *The Soviet Energy System: Resource Use and Policies* (New York: Halsted Press, Wiley, 1979), p. 253.

23 CIA Oil, July 1977, pp. 3, 32.

24 ibid., pp. 11, 23; *Review of Sino-Soviet Oil*, May 1977, pp. 4–5.

25 *Review of Sino-Soviet Oil*, April 1979, p. 13.

26 CIA Oil, July 1977, p. 12.

27 Feitel'man, 1978, p. 11.

28 Europa-Sibir', *EKO*, April 1976, p. 160; Dienes and Shabad, pp. 46–7.

29 CIA Oil, July 1977, p. 10; Dienes and Shabad, pp. 46–7.

30 CIA Oil, July 1977, p. 13.

31 ibid., pp. 8, 18; *Review of Sino-Soviet Oil*, December 1977, p. 15; *Soviet Business and Trade*, 1 August 1979, pp. 1–2.

32 *Review of Sino-Soviet Oil*, December 1977, pp. 15–16.

33 A. P. Krylov, 'O nekotorye voprosakh nefteotdachi', *Neftianoe khoziaistvo*, no. 3, 1974, p. 39; *Review of Sino-Soviet Oil*, December 1977, p. 16.

34 CIA Oil, July 1977, p. 11; *Sovetskaia Rossiia*, 28 February 1970, pp. 1–2; Dienes and Shabad, p. 58.

35 'Tiumen': Kompleke i ego grani', *Ekonomika i organizatsiia promyshlennogo proizvodstva*, no. 3, March 1979, p. 15 (hereafter referred to as Tiumen' *EKO*, March 1979).

36 *Review of Sino-Soviet Oil*, February 1979, p. 43.

37 *Pravda*, 1 April 1979, p. 1; *Review of Sino-Soviet Oil*, May 1979, p. 4.

38 Some Soviet economists are beginning to talk about using the future discounted value. Judith A. Thornton, 'Soviet methodology for the valuation of natural resources', *Journal of Comparative Economics*, no. 4, December 1978, p. 326.

39 Thornton, p. 330.
40 'Osnovnye problemy kompleknogo razvitiia zapadni sibiri materialy, kruglogo stola', *Voprosy filosofi*, September 1978, p. 35.
41 Campbell, 1976, pp. 16–17: CIA Oil, July 1977, p. 22.
42 *Current Digest of the Soviet Press*, 22 November 1978, p. 4; *Literaturnaia gazeta*, 18 January 1978, p. 10.
43 Tiumen' *EKO*, March 1979, p. 15; *Sotsialisticheskaia industriia*, 10 August 1979, p. 2; *Pravda*, 22 July 1979, p. 2, 26 July 1979, p. 2.
44 *Pravda*, 5 June 1978, p. 2; *Sotsialisticheskaia industriia*, 12 April 1978, p. 1.
45 *Current Digest of the Soviet Press*, 7 February 1979, p. 22, 14 March 1979, p. 20; *Soviet News*, 13 February 1979, p. 45.
46 *Current Digest of the Soviet Press*, 4 April 1979, p. 26, 7 February 1979, p. 22, 25 December 1978, p. 1, 15 April 1978, p. 9, 13 December 1978, p. 22; *Pravda*, 5 June 1978, p. 2.
47 *Current Digest of the Soviet Press*, 20 June 1979, p. 8.
48 *Pravda*, 20 April 1978, p. 2; *Sotsialisticheskaia industriia*, 17 May 1978, p. 2, 21 August 1979, p. 2; *Current Digest of the Soviet Press*, 10 November 1977, p. 18.
49 *Wall Street Journal*, 26 October 1978, p. 48.
50 *Pravda*, 5 June 1978, p. 2; B. Vainshtein, A. Khaitun and N. Sokolov, 'Effektivnost' neftegazovogo kompleksa zapadnoi Sibirii', *Voprosy ekonomiki*, October 1979, p. 27.
51 *Pravda*, 29 December 1972, p. 1, 9 July 1975, p. 1; *New York Times*, 21 January 1974, p. 1; *Izvestiia*, 20 August 1976, p. 2, 30 July 1977, p. 3.
52 *Pravda*, 14 June 1979, p. 1.
53 *Sotsialisticheskaia industriia*, 31 January 1979, p. 2.
54 Tiumen' *EKO*, March 1979, p. 10.
55 *Sotsialisticheskaia industriia*, 31 January 1979, p. 2.
56 *Current Digest of the Soviet Press*, 13 January 1979, p. 5.
57 *Izvestiia*, 2 December 1978, p. 5; *Review of Sino-Soviet Oil*, February 1979, p. 44; *Pravda*, 16 August 1979, p. 3.
58 A personal communication with the American manufacturer.
59 *Sotsialisticheskaia industriia*, 2 August 1979, p. 2.
60 *Wall Street Journal*, 5 April 1979, p. 28; *Soviet Business and Trade*, 1 August 1979, pp. 1–2.
61 *Moscow Narodny Bank Bulletin*, 2 November 1977, p. 7; *Soviet Business and Trade*, 15 August 1979, p. 3.
62 *Moscow Narodny Bank Bulletin*, 5 April 1978, p. 9; *Review of Sino-Soviet Oil*, April 1977, pp. 48–9.
63 ibid., August 1979, p. 65.
64 *Business Week*, 14 June 1976, p. 31; *Energy Insider, Department of Energy*, Washington, DC, 30 April 1979, p. 3.
65 *Current Digest of the Soviet Press*, 8 February 1978, p. 1.
66 CIA, August 1979, p. 39.
67 *Review of Sino-Soviet Oil*, August 1979, pp. 1–7.
68 L. A. Potemkin, *Okhrana nedr i okruzhaiushchei prirody* (Moscow: Nedra, 1977), p. 107. Work in Tiumen has also been reorganized. *Sotsialisticheskaia industriia*, 2 August 1979, p. 2.
69 *Review of Sino-Soviet Oil*, March 1978, p. 53.
70 *Current Digest of the Soviet Press*, 2 May 1979, p. 2; *Review of Sino-Soviet Oil*, December 1978, p. 28.
71 ibid., August 1979, p. 1.
72 *Ekonomicheskaia gazeta*, no. 35, August 1979, p. 2.

73 *Soviet Business and Trade*, 1 August 1979, pp. 1–2; *Moscow Narodny Bank Bulletin*, 1 August 1979, p. 2.

74 *Oil and Gas Journal*, 1 September 1975, p. 63.

75 V. I. Muravlenko and V. I. Kremneva (eds), *Sibirskaia Neft'* (Moscow: Nedra, 1977), p. 18; American Petroleum Institute, *Petroleum in the Soviet Union* (Washington, DC: mimeo., undated), p. 60; Elliott, p. 81; J. Richard Lee, 'The Soviet petroleum industry: promise and problems', *Soviet Economic Prospects for the Seventies*, Joint Economic Committee, Congress of the United States (Washington, DC: US Government Printing Office, 1973), p. 283.

76 *Total Information*, no. 68, 1976, p. 10.

77 *Soviet Business and Trade*, vol. III, no. 5, 18 July 1979, p. 1.

78 *Review of Sino-Soviet Oil*, January 1978, p. 26; *Oil and Gas Journal*, 18 September 1978, p. 66.

79 *Review of Sino-Soviet Oil*, April 1979, p. 11.

80 ibid., April 1979, pp. 4, 10, 11.

81 ibid., April 1979, p. 74, July 1979, p. 60.

82 *Voprosy filosofi*, September 1978, p. 35; *Sotsialisticheskaia industriia*, 2 August 1979, p. 2.

83 *Nedelia*, no. 49, December 1978, p. 2; *Sotsialisticheskaia industriia*, 2 September 1978, p. 2.

84 *Oil and Gas Journal*, 1 September 1975, p. 63; Muravlenko and Kremneva, p. 18.

85 *Sovetskaia Rossiia*, 8 December 1978, p. 2.

86 *Nedelia*, no. 49, December 1978, p. 2; *Current Digest of the Soviet Press*, 19 May 1979, p. 9.

87 *Nedelia*, no. 49, December 1978, p. 2; *Petroleum Economist*, December 1978, p. 506; *Sotsialisticheskaia industriia*, 15 September 1978, p. 1, 20 October 1978, p. 2; *Literaturnaia gazeta*, 18 January 1978, p. 10; *Pravda*, 19 February 1975, p. 3, 14 July 1978, p. 2, 26 July 1979, p. 2; *Review of Sino-Soviet Oil*, June 1977, p. 18, December 1977, p. 37; *Petroleum Economist*, August 1978, pp. 3, 48; *Soviet News*, 21 December 1976, p. 445; *Current Digest of the Soviet Press*, 22 November 1978, p. 7; *Oil and Gas Journal*, 25 September 1978, p. 65; Dienes and Shabad, pp. 60–1.

88 *Moscow Narodny Bank Bulletin*, 19 October 1977, p. 5.

89 *Review of Sino-Soviet Oil*, September 1976, p. 56.

90 *Moscow Narodny Bank*, 12 April 1978, p. 10; *Review of Sino-Soviet Oil*, June 1979, p. 46.

91 *Sotsialisticheskaia industriia*, 27 September 1978, p. 3; *Review of Sino-Soviet Oil*, October 1978, p. 33.

92 ibid., April 1979, p. 12.

93 *Ekonomicheskaia gazeta*, no. 42, October 1978, p. 11; Iu. Iakovets, 'Ekonomicheskie rychagi i povyshenie effektivnosti minerali 'no-syr' evogo kompleksa', *Planovoe khoziaistvo*, January 1978, p. 69; Robert G. Jensen, (ed.), *Soviet Energy Policy and the Hydrocarbons: Comments and Rejoinders* (Washington, DC: American Association of Geographers, February 1979), no. 7, p. 31.

94 Nekrasov and Pervukhin, p. 74.

95 Iakovets, 1975, p. 8; Lalaiants, p. 28; *Sotsialisticheskaia industriia*, 10 August 1979, p. 2; *Soviet Weekly*, 7 July 1979, p. 5.

96 Levin, Vasil'ev and Kosinov, p. 110; *Problems of Economics*, April 1978, p. 72.

97 Nekrasov and Pervukhin, p. 144.

98 ibid., p. 149.

99 CIA, 1975, p. 15; A. Probst, 'Puti razvitiia toplivnogo khoziaistva SSSR', *Voprosy ekonomiki*, June 1971, p. 59; N. S. Sazykin, 'Mineral 'no-syr' evye resursy', *Znanie seriia nauka o zemble* (Moscow), June 1975, p. 29; Ia. Mazover, 'Puti razvitiia toplivnogo khoziaistva SSSR', *Planovoe khoziaistvo*, November 1977, p. 66; Ia. Mazover, 'Perspektivy Kansko-Achinskogo ugol'nogo basseina', *Planovoe khoziaistvo*, June 1975, p. 42.

100 Ia. Mazover, 1977, p. 146; Russell, p. 377; A. Beschinskii and R. Vitebskii, 'Energetika i razmeshchenie promyshlennogo proizvodstva', *Voprosy ekonomiki*, May 1977, p. 44; D. V. Belorusov, I. I. Panfilov and V. A. Sennikov, *Problemy razvitiia i razmeshcheniia proizvoditel'nykh sil zapadnoi Sibiri* (Moscow: Mysl', 1976), p. 257.

101 Dienes and Shabad, p. 118; Campbell, 1978, pp. 42–5.

102 *Soviet Weekly*, 21 July 1979, p. 7; *Soviet News*, 15 May 1979, p. 151.

103 *Ekonomicheskaia gazeta*, no 41, October 1978, p. 2.

104 *Pravda*, 24 September 1978, p. 2.

105 Iakovets, 1975, p. 6; A. Troitskii, 'Novye rubeshi Sovetskoi energetiki', *Planovoe khoziaistvo*, December 1976, p. 44.

106 Levin, Vasil'ev and Kosinov, p. 111; *Sotsialisticheskaia industriia*, 16 August 1979, p. 2.

107 *Nar. khoz.*, 1977, p. 544.

108 Probst, p. 53.

109 Mazover, 1975, p. 66; *Total Information*, no. 66, 1976, p. 9; Russell, p. 37.

110 *ASTE Bulletin*, vol. XII, no. 2, Fall 1970, p. 11; Iakovets, 1975, p. 9.

111 Iakovets, 1975, p. 9; Vainshtein, Khaitun and Sokolov, p. 22.

112 *New York Times*, 7 August 1979, p. D14.

113 *Nar. khoz.*, 1977, p. 143.

114 ibid., pp. 543–4.

115 Vainshtein, Khaitun and Sokolov, pp. 22–3; Evropa-Sibir', *EKO*, p. 171; *Current Digest of the Soviet Press*, 7 March 1979, p. 11.

116 *ASTE Bulletin*, vol. XII, no. 2, Fall 1970, pp. 11–13.

7

Conservation, Substitution and Export Policy – American and Russian

While the Soviets obviously would like to sustain petroleum production growth, they have other options if production should fall. These options include conservation, particularly of petroleum, and the substitution of other energy sources for petroleum. Even if petroleum production declines, the Soviets might be able to continue petroleum exports.

FUEL BALANCE AND ENERGY CONSERVATION

Because of peculiarities in the Soviet planning and price system, energy consumption patterns in the Soviet Union are quite different from those in the non-communist world. Unlike the United States, the Soviets have a relatively small stock of automobiles, the lowest per capita of any industrialized society. In addition, the Soviet public transportation system is one of the world's most effective. Therefore, economizing on unneccessary automobile and truck travel in the Soviet Union will not produce the savings it would in a country like the United States. Whereas the 130 million or so cars and trucks in the United States consume as much as one-third of total United States energy, Robert Campbell estimates that the 18 to 20 million automotive vehicles in the Soviet Union consume only about 10 per cent of that country's total energy.[1] (See Table 7.1.) But that does not mean that the Soviets have no margin for conservation. They doubled the price of gasoline in early 1978, a rather draconic step for a society that makes a big point of keeping its retail prices constant. Of course, having access to gasoline coupons is more important in determining the sales of gasoline in the Soviet Union than having access to money. But there were also signs that in the high-export years of 1976 and 1977 the Soviets were also curbing the allocation of gasoline. They also appear to have made a conscious effort to curb automobile output. Whereas in the Ninth Five-Year Plan from 1970 to 1975 the Soviets increased automotive vehicle (trucks, cars, buses) output by a phenomenal 130 per cent, in the Tenth Five-Year Plan automotive vehicle production is increasing by only

Table 7.1 Net Energy Consumption by Final Demand Sector* (quadrillion BTUs per year)

	Area	1975
Total consumption**	USSR	27·3
	USA	47·9

Demand Sector's Share of Consumption (%)

Demand Sector	Area	1975
Industry	USSR	59
	USA	35
Transportation	USSR	10
	USA	34
Other – Residential, Commercial, Construction, Public Use, and Agriculture	USSR	32
	USA	31

* 'Net energy consumption' is used here in the sense used by the US Bureau of Mines, i.e. to include the net energy value of electricity consumed in addition to gross direct energy use. In the Soviet case, it also includes by-product heat from co-generation.
** In some cases components do not add to totals because of rounding.

Source:
Campbell, *Soviet Energy Balance*, p. 22.

7–12 per cent. In contrast to the Ninth Five-Year Plan increase of automobile output by 350 per cent, the new plan does not even contain a figure for automobile production. Interviews with the director of the new Fiat-built automotive plant at Togliatti reveal his frustrations at not being able to increase production more than is authorized by the Tenth Five-Year Plan.

In contrast to transportation, more Soviet energy is allocated to residential use. As in the United States, energy for Soviet residential use accounts for about one-third of all energy consumed. (See Table 7.1.) In many respects, their urban heating system is efficient.[2] They have a system of regional heating that obviates the need for a furnace and hot water heater in each apartment house. Since much of this heat is co-generated with electricity, it also serves to utilize what otherwise would be regarded as thermo-pollution. Yet, there is room to increase energy efficiency even here. At present, not all Soviet housing units are supplied this way. The goal is to have 45 per cent of Soviet urban homes heated with co-generated heat by 1980.[3] We in the United States are just beginning to design such systems.

Yet, while the Soviets may spend less energy on household heat per degree day and on per capita use of automobiles, there are sectors where energy is squandered.[4] For example, building insulation is poor and drafts are endemic. S. Orudzhev, the Minister of the Gas Industry, complained that it takes 60 per cent more fuel to heat buildings erected in the 1970s than those built twenty-five years earlier.[5] If better insulation were to be used, savings of 15 billion cubic meters of gas a year could be made. In other cases, centrally heated buildings are often too hot. Operators of heating plants are evaluated on the basis of how much heat they generate, not how much energy they save.[6] Since Soviet radiators do not always function as advertised, and most do not have individual control valves, many residents are forced to open the windows if they want to cool their apartments, even in the middle of winter.[7] Because of sloppy work habits, much waste also takes place in Soviet factories, particularly in metallurgical and ore refining enterprises. Similarly, because drivers in most of the Soviet Union's motor pools are rewarded premiums, not for the tasks performed, but on the kilometers driven, drivers both in the cities and on the farms are constantly criticized for wasting gasoline. This may explain the incredible sight of herds of elephant-like water trucks roaming Moscow streets with spray spewing from their trunks even in the rain. Other drivers tamper with their speedometers, showing that they have covered the assigned distances, and then illegally sell their gas on the black market.[8]

Discouraging and even frustrating as such waste may be, most of the Soviet problems discussed thus far stem from human or systemic failings. They may be difficult to correct, but not impossible. For example, the CIA finds that Soviet energy consumption per unit of GNP, after increasing by 1 per cent a year from 1971 to 1976, leveled off in 1977.[9] Calculations by Soviet authorities indicate the same findings.[10]

The Soviets feel they can continue this conservation. Undoubtedly, fuel conservation would be improved if prices were more reflective of the opportunity cost between various fuels in the Soviet Union and between domestic and export use. Few Soviet economists dare to point this out. N. Tikhonov, a candidate member of the Politbureau and the first deputy president of the Council of Ministers, at least recognizes the problem.[11] Those with lesser positions restrict themselves to observing that energy resources are not always used in the most effective way.[12]

Some look to increased efficiency in industry for the largest and most necessary energy savings. The Ministries of Energy, Heavy Machine Building, Ferrous Metals, Construction Materials, and Chemical Construction consume 60 per cent of all the country's fuel.[13] In contrast, industry in the United States consumes about 35 per cent of our total energy. (See Table 7.1.) Since some of the more efficient Soviet enterprises utilize 15–20 per cent less fuel than the average, there is reason to believe that savings of 10–15 per cent are possible in almost all of these ministries.[14] Obviously this will not be easy, particularly since the Soviet price system, which undervalues domestic energy, cannot be depended upon to induce enterprise managers to act on their own. Yet in some ways it is less difficult to pressure a few big industries than millions of individual households and drivers, as in the United States.

While strictly speaking it is not conservation, Soviet planners are also trying earnestly to substitute coal, gas, and nuclear power for petroleum. In the past these efforts have not always been carried out rationally. When an order was issued to eliminate oil as a source of power generation, local officials in one instance carried out the order even though they had nothing but abundant quantities of petroleum immediately available to them. To carry out the order they had to ship coal hundreds of miles, thereby wasting more energy in the process.[15]

Recently, however, conservation of petroleum for internal and external purposes has been carried out in a somewhat more rational manner. As the United States has discovered, this is not easy. The switch away from petroleum is dependent on more than a sudden quadrupling or doubling of the price. Over the course of thirty years or so, the move from coal to petroleum and other energy resources in the United States generated and in turn was generated by basic structural changes in our personal and economic lives. Attracted by cheap gasoline, the United States in the 1950s of the Eisenhower years embarked on a far-reaching interstate highway construction program. The effect of this program was much more dramatic than we imagined at that time. A good highway system made it easier for the white middle class to move out of the central city. Initially the moves followed commuter railroad lines, but eventually new suburbs sprung up without regard to public transportation. What had been a reasonably efficient public transportation system began to disintegrate. After a time, there were not enough riders. As large numbers of their former customers relocated to the suburbs, urban transportation systems came under the same financial pressures. Gradually city factories and even offices also moved out to the suburbs chasing their employees. Once these factories and offices were built in beautiful but decentralized locations, it became even more difficult to sustain a public transportation system that depended on a heavily centralized concentration of population and work destinations to be viable. Sometimes even massive federal subsidies could not help. To add insult to insanity, the gasoline tax revenues, which might have been used to subsidize public transportation, were restricted to a highway trust fund. Until the early 1970s the fund's proceeds could only be used to build more highways.

The whole process evolved over a quarter of a century. Each year as many as 10 million automobiles were added to the nation's highways along with billions of dollars' worth of new construction designed to accommodate the new automotive way of life. Inevitably this new way of life gave rise to a radical displacement, a revolution if you will, in white middle-class living patterns. Once altered, what had taken twenty-five years to tear down and build up could not be restored overnight. Thus, when a fourfold price hike in the price of petroleum in 1973 and a twofold price hike in 1979 with long gas lines in both instances made people realize that they had become too dependent on the automobile, there were no quick fixes. Just as it took twenty-five years to get into this mess, so it may take almost as long to get out of it, and even with twenty-five years to do it, the adjustment is not likely to be tranquil. A shift to small cars helps, but that is more like a band-aid when it is radical surgery that is called for. Any basic solution necessitates an undoing of the suburbs and a regathering of people, offices, and factories into central locations. Only then will it be possible to return to any massive use of public transportation.

The shift from cities to suburbs was only a part of the process. Virtually the entire economy was affected by the same type of transformation. In all instances, the shift was predicted on extravagant use of cheap petroleum. Regions of the country that formerly were regarded as uncomfortable because of the heat and cold became year-round resort areas through the miracle of air-conditioning and indoor malls. Houston and New Orleans with their enclosed stadiums attracted large populations that previously could not stand the stifling summers. Skyscrapers with their sealed glass windows only made sense if there was cheap energy. Railroad rights of way were abandoned, and factories manufacturing railroad and transportation equipment were closed down for lack of business. Finally, electric power generating plants, especially along major water transportation routes, closed down their dirty coal yards and in their place built oil storage tanks. The reduction in the use of coal not only reduced air pollution, but it also allowed the utilities to sell off valuable properties, because oil storage took less space than coal storage.

The earlier a society begins to switch away from coal-generated energy, the more radical and final that switch is likely to be, and the harder it will be to give up petroleum. Once they are out there, how can you move people around the suburbs without cars? What will all the people in the sun belt do as air-conditioning becomes too expensive? What can be done with the thousands of skyscrapers that have no insulation but plenty of sealed windows? How do you restore the railroads now that some of the roadbeds have been sold off? One consolation is that the shock in 1973 came as early as it did. If it had come later, the pain of the readjustment would be even more traumatic than it is now.

Ironically the readjustment may be less difficult for the Soviet Union. Few thought so at the time, but its inflexible pricing and planning system protected it from the corrupting influence of cheap petroleum long after almost everyone had succumbed in the West. Remember that in the USSR the relative consumption of petroleum exceeded the relative consumption of coal only in 1973. Thus the Soviets have not had enough time to destroy their center cities and move out into the suburbs. Some upper-class Soviet officials have started to use their cars to commute from their dachas in the summer, but that is a very small percentage of the total population. Even that 5 per cent of the Soviet population that have cars generally use public transportation to commute to their jobs. Personal automobiles in the Soviet Union are used primarily for weekend visits and trips rather than for daily transportation. Nor have the Soviets moved yet to a massive highway construction program. There were hints of a move in that direction, particularly as the new Togliatti Fiat-type car and the Kama River trucks started coming off the assembly line, but the retardation in automobile vehicle production in the mid-1970s has reduced the urgency of this ultimately corrosive influence. Thus, Soviet public transportation remains relatively healthy.

Nor have the Soviets become so dependent on petroleum that they have forgotton how to use coal and other fuels. They have not had time yet to plough up their coal storage yards, dismantle the coal-moving infrastructure, or convert all their boilers to petroleum as we have so often done in the United States.[16] Moreover, they acted early to halt the drift to petroleum. In January 1975 they announced that coal should replace the use of petroleum wherever possible.[17] That was particularly important in the Ural and Volga regions. However, it is not easy to make mid-course corrections of this sort, so it will take until at least 1980 before the relative importance of coal in boiler use and in the fuel balance in general will cease to diminish.[18] (See Table 3.1.) In the far north-east where coal is unavailable, and in west Siberia where gas is abundant, the second-best solution is to burn gas. Indeed, it is anticipated that the use of natural gas will increase significantly both absolutely and relatively. Thus its use as a source of boiler and furnace fuel is slated to grow from 29·6 per cent in 1975 to 34·8 per cent in 1980. Natural gas, and to a small extent liquified gas, are the only fuels whose use with boilers is due to increase over the five-year period. (See Table 3.1.) Natural gas as a percentage of overall energy production is to increase from 21·2 to 24·5 per cent. (See Table 3.3.) Apparently gas will even replace wood as a fuel, although timber will continue to be relatively important.

The Soviets are particularly proud of their effort to use previously flared gas from Nizhnevartovsk at Surgut, Kemerovo, and Novokuznetsk as fuel in boilers or in thermo-electric generating plants and even in steel mills.[19] The piping of the gas to the Kuznetsk Basin will result in the saving of at least 2 million tons a year of coking coal.[20]

GAS

The Soviets are fortunate in that their reserves of gas and coal are extensive. In the case of gas, the Soviets have the largest reserves in the world. Depending on which calculations are used, the Soviet Union contains 35 to 40 per cent of all the natural gas reserves in the world.[21] The reserves at Urengoi, the world's largest gas field, are alone estimated to total as much as 5 trillion cubic meters of gas.[22] Supplementing this are four other giant fields at Yamburg, Zapoliarnoe, Medvezhye, and Orenburg. Each field has reserves of more than 1 trillion cubic meters. Another large field exists in Turkmenistan in Central Asia, where a large new field was confirmed in 1978.[23] Because of these abundant supplies the Soviets do not regard their estimates of natural gas reserves as a state secret. According to official reports, natural gas reserves in the Soviet categories $A + B + C_1$ rose from 22·5 trillion cubic meters (TCM) in 1974 to 23 TCM in 1975 to 29 TCM in 1978.[24] Some foreign estimates are more conservative and place reserves at 632 trillion cubic feet or about 18 TCM.[25] Estimates of future potential reserves range from 85 to 150 TCM (3,016 to 5,300 trillion cubic feet).[26] This includes the already proved reserves of 29 TCM mentioned earlier.

Despite these enormous reserves, however, increasingly the more accessible fields in the Ukraine and the northern Caucasus are being depleted. While new discoveries in Turkmenistan and Uzbekistan are welcome, there is concern, especially by consumers in Uzbekistan, that other important fields in their republic are being depleted.[27] Indeed, there is a good deal of regional jealousy and resentment about the export of energy reserves from one region of the country to another.[28]

At one time depletion of these fields in the warmer climate caused considerable alarm.[29] Because no one had experience in constructing pipelines in the permafrost areas, enormous and costly logistical problems had to be solved.[30] But as elsewhere, most of them eventually were solved. Moreover, the deposits were so large that the ultimate delivered cost of the gas was not as high as first feared. Yet, because it requires five times more natural gas than oil to pipe an equivalent amount of heat energy, the Soviets are trying to use coal in place of both gas and oil. In this way they hope to conserve oil and gas for higher value uses, including exports. If necessary, however, gas will be used domestically to replace oil, because it is cleaner and more abundant. Also, if electricity can be provided and petrochemical plants can be built at sites near where gas is produced, such as Surgut, Nizhnevartovsk, Tobolsk, and Tomsk, then the piping cost can be radically reduced.[31] In the same way, gas has been converted into ammonia and an ammonia pipeline network has been constructed to carry the ammonia westward to Odessa. Along the way some of it is converted into fertilizer for domestic use. The remainder is exported and ultimately will become an important hard-currency earner. For example, Occidental Petroleum has agreed to import over 1 million tons of ammonia a year into the United States, or it will assuming that Occidental is able to diffuse the opposition of American ammonia manufacturers who see Soviet ammonia as a competitive threat.

COAL

Still, wherever possible, the emphasis remains on coal.[32] Soviet coal reserves are not as extensive or tractable as Soviet gas supplies, but the quantity available is still enormous. The CIA estimates that Soviet reserves total 109·9 billion tons, second only to the United States with 178·6 billion tons.[33] The Soviets estimate, however, that reserves of $A + B + C_1$ deposits amount to 255–76 billion tons. This includes 108 billion tons of lignite.[34] If in addition, categories C_2 and D (Prospective) reserves are included, then the Soviet Union is estimated to have 59 per cent of the world's reserves.[35]

While the potential is there and while they have not forgotten how to burn coal, the Soviets have experienced difficulties in increasing their absolute production of coal, not to mention its relative production. The target set for the Tenth Five-Year Plan has been badly underfulfilled. Output in 1980 was projected at 805 million tons which required an increase in production over the five years of 104 million tons. However, instead of average increments of about 20 million tons a year, the Soviets seldom were able to attain 10 million tons. In 1978, production only increased by about 2 million tons. The traditional underground mines in the west have become more and more expensive to operate, so that despite the decision to upgrade coal, some of them were closed down.[36] At the same time, work at the newer and most promising fields where coal is near the surface, and can therefore be strip-mined, is moving slowly because of impurities in the coal and the higher investment required to make these fields operable.[37] Given these problems, the Soviets may not be able to free up as much petroleum and gas as they anticipated. The solution will probably be to use more gas domestically than preferred. They also are moving ahead determinedly with nuclear energy as a substitute for coal as well as with oil and gas.

NUCLEAR ENERGY

Nuclear energy is expected to play a particularly important role in the populated area of the western part of the Soviet Union. These older and more heavily industrialized regions of the country are almost all energy-short. The Soviets are especially interested in utilizing nuclear energy for electrical generation and eventually household heating. While they were the first in the world to build a nuclear generating plant on 27 June 1954, at Obminsk, once the plant was up they did not do much more with nuclear energy until the 1970s.[38] As long as other Soviet fuels were so cheap and abundant, it was not a pressing matter.[39] But as conventional reserves in the west became depleted, the Soviets began a rush program to increase the number of nuclear energy facilities. With one exception, all of these facilities were to be concentrated in the European part of the Soviet Union. Philip Pryde, in one of the best studies written on Soviet nuclear energy, has compiled a list of nuclear sites already completed and under construction. (See Table 7.2.)

While the overall share of nuclear energy will continue to be small, the share of nuclear energy in the European part of the Soviet Union is due to increase from 3·1 per cent of all energy generated in the western Soviet Union in 1975 to 10 per cent in 1980.[40] During the five-year period from 1976 to 1980, 35 per cent of all the new energy capacity to be opened in the European part of the Soviet Union is supposed to be produced by nuclear reactors. However, the plan is running behind

Table 7.2 Nuclear Power Reactors in the USSR

Station name and reactor number	Station site	First Power output	Commercial MW
Obninsk	Obninsk	1954	5
Siberian	Troitsk	1958	600
Obninsk	Obninsk	1959	5
Beloyarskiy-1	Zarechnyy	1964	100
Novovoronezhskiy-1	Novovoronezhskiy	1964	210
Dimitrovgrad	Dimitrovgrad	1965	50
Beloyarskiy-2	Zarechnyy	1967	200
Novovoronezhskiy-2	Novovoronezhskiy	1969	365
Dimitrovgrad	Dimitrovgrad	1969	12
Novovoronezhskiy-3	Novovoronezhskiy	1971	440
Novovoronezhskiy-4	Novovoronezhskiy	1972–3	440
Shevchenko	Shevchenko	1973	150
Bilibino-1, 2, 3	Bilibino	1973–5	36
Kola-1	Polyarnyye Zori	1973	440
Leningrad-1	Sosnovyy Bor	1973	1,000
Kola-2	Polyarnyye Zori	1974	440
Leningrad-2	Sosnovyy Bor	1975	1,000
Bilibino-4	Bilibino	1976	12
Kursk-1	Kurchatov	1976	1,000
Armenia-1	Metsamor	1976	405

Chernobyl'-1	Pripyat'	1977
Chernobyl'-2	Pripyat'	1978
Kursk-2	Kurchatov	1979
Novovoronezhskiy-5	Novovoronezhskiy	(1979)
Leningrad-3	Sosnovyy Bor	(1979)
Rovno-1	Kiznetsovsk	440	(1979)
Beloyarskiy-3	Zarechny	600	(1979)
Armenia-2	Metsamor	410	(1979)
Kursk-3	Kurchatov	1,000	(1980)
Leningrad-4	Sosnovyy Bor	1,000	(1980)
Rovno-2	Kuznetsovsk	440	(1980)
South Ukraine-1	Konstantinovka	1,000	(1980)
Smolensk-1	Desnogorsk	1,000	(1980)
Kola-3	Polyarnyye Zori	440	(1980)
Kola-4	Polyarnyye Zori	440	(1980)
Kalinin-1, 2, 3, 4	Udomlya	1,000 each	(after 1980)
West Ukraine-1, 2, 3, 4	Khmel'nitskiy	1,000 each	(after 1980)
South Ukraine-2, 3, 4	Konstantinovka	1,000 each	(after 1980)
Kursk-4	Kurchatov	1,000	(after 1980)
Chernobyl'-3, 4	Pripyat'	1,000 each	(after 1980)
Smolensk-2	Desnogorsk	1,000	(after 1980)
Ignalina-1, 2	Sneckus	1,500 each	(after 1980)

Source:
Philip Pryde, in Dienes and Shabad, *Nuclear News*, August 1975, pp. 74–5.

schedule.[41] In 1979 Soviet capacity amounted to 10,500 megawatts, and nuclear energy output amounted to 4·4 per cent of all electrical output. The Tenth Five-Year Plan called for an increase of 13,200 megawatts capacity by 1980 with nuclear energy accounting for almost 6 per cent of the total electricity produced.[42] Yet once the Soviets complete construction on their reactor factory called Atommash at Volgadonsk, construction of nuclear energy plants should accelerate.[43]

The plant at Volgadonsk supplemented by the reconstruction of the Izhorski plant in Leningrad should make it possible for the Soviets to add as much as 8 million kilowatts of capacity a year by the late 1980s and as much as 10 million kilowatts a year by the late 1990s. These plants will build pressurized water vessel reactors, a type more common to the west. The Soviets also have been building a 'channel type' reactor that does not require specialized engineering facilities. 'Channel type' reactors are water cooled and have a graphite moderated reactor. They also have individual fuel channels so that, if one channel breaks down, it can be shut down without affecting the other operating channels.

The Soviets have not limited themselves to ordinary fission reactors. They have done extensive work with breeder reactors as well. The first breeder reactor began operating at Shevchenko on the east shore of the Caspian Sea in 1972, and another is being built at Beloyarskii.[44] They have also committed themselves to advanced experiments with thermo-nuclear fusion. So far much remains to be done, but the Soviets hope that by 1990 the first thermo-nuclear power stations will be built.[45]

The Soviets have a very important advantage in moving toward nuclear energy that is lacking in almost all the other countries of the world except France. They are unhampered by environmental or safety protests about nuclear energy operations or waste disposal. Unlike Western officials, who insist that nuclear energy plants be located far outside populated centers, Soviet planners have now decided that they will build some of their new nuclear plants in the midst of residential districts.[46] The plants, then, can be used not only to generate electricity but to co-generate steam for household heating that can further reduce the need to burn conventional fuels.[47] An experimental unit involving co-generation of nuclear energy and heat was built in 1978 at Bilibino in Siberia and located 3 miles from the center of the town. Similar plants are under construction in Gorky and Voronezh.[48]

Although Robert Campbell has found some concern, Soviet scientists normally admit to few fears about such procedures.[49] And those fears tend to focus on nuclear waste processing procedures more than anything else. For that matter, until the Three Mile Island incident in Pennsylvania in 1979, the Soviets did not even bother to build containment shells round their reactors. In their minds, reactor safety is not a problem. As Anatoly Alexandrov, the president of the Academy of

Sciences, sees it, such facilities 'can be built in the middle of residential areas, because they are absolutely safe'.[50] Presumably Soviet scientists would not build nuclear energy plants if they were not safe. Thus, the Soviet public is expected to believe that nuclear energy is safe; it is anti-nuclear protests that are suspect. 'The actual reason behind the whole fuss over nuclear construction in the United States', according to Alexandrov, 'has nothing to do with safety. The real reason is that the development of large nuclear power stations could endanger the profits of the fuel producing monopolies'. This is said despite the fact that Pyotr Neporozhny, the Soviet Minister of Power and Electrification, has conceded that accidents have occurred at Soviet nuclear plants. Several unofficial observers have pointed to a major nuclear catastrophe near the city of Kyshteym in the Urals.[51]

While we may not like Alexandrov's reasoning, the point is that the Soviets do have an energy policy. The first priority is to increase the relative share of energy generated by nuclear power as opposed to conventional fuels. They recognize, however, that even if all their power stations were converted to nuclear energy, they would reduce the consumption of regular fuel by only 20 per cent.[52] Furthermore, since many of the existing power stations are fueled by coal, switching to nuclear energy would probably only reduce oil and gas consumption by 10 per cent at the most. None the less, Harvey Brooks of Harvard University points out that, when the Volgadonsk and Czechoslovak nuclear reactor plants each begin producing eight to ten nuclear reactors a year, these reactors when installed and operating will displace about 0·5 to 0·6 MBD of petroleum a year. Thus, if the Soviets are able to bring on line a new set of reactors each year, after ten years, nuclear power alone will compensate for the 4·5 MBD of CMEA petroleum imports predicted by the CIA.

But it is not only a question of reactor capacity. Concern has also been expressed that the Soviet Union may find it increasingly difficult to obtain the uranium deposits it needs.[53] Supposedly these fears will subside if, as rumored, vast new uranium deposits are opened in the Ukraine. Such deposits would free the Soviets from dependence on Eastern European supplies and assure the Soviets of self-sufficiency for some years to come.[54] Yet, even then, nuclear energy could not supply the bulk of Soviet energy needs. Therefore other means must be found to reduce oil and gas consumption. Although some of the bigger rivers are already dammed up, and most of the potential is in Siberia, something more can be done with water. The Soviets are also experimenting with a process called MHD (Magnetohydrodynamics), which is a means of reducing inefficiencies in the production of energy. They are also trying to increase their use of solar and geothermal energy.[55] But of all the alternatives, an increase in the use of coal and gas and an increase in energy efficiency are the most promising approaches.

While the Soviets are capable of resolute action when it comes to the construction of new facilities, ironically even they have their frustrations. True, they are spared most of the delay caused by environmental groups in this country, nor must they concern themselves with complaints that their actions will bring enormous windfall profits. Still, the Soviets find they can move only so far in raising prices. They increased the wholesale price of gasoline in 1967 and doubled the retail price in 1978, but otherwise they have been very cautious. Price stability in the Soviet Union is a sacred principle. Price increases in the past have been the cause of popular protest in the USSR as well as in Eastern Europe. Hence the Soviets show more restraint than is economically sensible. But because physical allocation is normally a more crucial factor in the distribution of goods than demand elasticity, this is not a crucial shortcoming. When goods are in short supply, or the Soviets find it necessary to decrease the amount set aside for consumption in order to increase exports, they simply cut off domestic deliveries. It would be economically more efficient, perhaps, if they could rely on demand adjustments to price changes to facilitate that process, but supply adjustments will do if necessary. But whether the Soviets concentrate on increasing supplies or restraining demand or both, it is important to recognize that they have a number of viable options.

WHAT IF THE CIA IS CORRECT?

Despite all the alternatives open to the USSR, let us none the less assume that the CIA is correct and that instead of being just a 'notational gap', the Soviets and their East European allies find themselves forced to import 2·7 to 4·5 MBD of petroleum? What does that imply for the Soviet economy as well as for its foreign policy?

For the domestic economy, imports of this magnitude would be disastrous. If, as predicted in Chapter 6, the Soviets have to spend $25 billion to $30 billion in hard currency for their purchases, the Soviet Union would literally have no other money left to import anything else including grain. Inability to import foreign products would cause enormous harm and make it all but impossible for the Soviets to use imports to eliminate bottlenecks as they now try to do. Anything approaching shortages of this magnitude would spell chaos for the Soviet and East European economies. The energy-intensive industries which predominate in the CMEA would not only become unprofitable, it may be that there would not be enough energy to sustain their operation. As a minimum, there would be an enormous decline in the

GNP of each of the members of the CMEA, which in the West we would call a depression.

If such a shortage of energy were to threaten their domestic economies in this way, the Soviet Union and Eastern Europe might act to secure outside sources of supply regardless of whether or not they had hard currency. They have arranged this in the past through the barter of petroleum or gas for industrial and military aid. Major recipients have been Iran, Libya, Iraq, Syria, Afganistan, Egypt, and Algeria. When an Arab country like Iraq decides that the Soviet Union must pay for its petroleum with hard currency, the Soviets agree but then in response charge Iraq hard currency for Soviet arms, so that generally no hard currency is transferred.

The big unknown is whether the arms traffic will be large enough to generate the billions of dollars needed annually to import as much oil as the CIA predicts. Certainly these sales will help, but it hardly seems likely that they will be large enough to satisfy the total petroleum import bill. In that event, particularly since it is doubtful that anyone will lend the Soviets and Eastern Europeans the billions of dollars they will need each year to pay their petroleum bill, the Soviets may find themselves impelled to take some political or military action.

One alternative would be to invade some of the Middle East oil producing countries. Given the Soviet invasion of Afghanistan, that is no longer the unlikely possibility that it once seemed to be. Yet given the rather expected strong world reaction to the Soviet move into Afghanistan, it seems likely that the invasion of an oil producing country would arouse an even greater storm and might conceivably be one of the few moves that would rouse the West to counter military action. Consequently, a more likely scenario would be to stir up political trouble in the hope that it would lead to a revolution that in turn could lead to the intervention of their allies like Cuba, the East Germans, or the Vietnamese.

Politically the Soviet Union would consider it a victory if a relatively pro-Western country like Saudi Arabia turned neutral. The Soviet Union does not necessarily seek subservience from such countries. Yet, welcome as antagonism or even neutrality toward the West might be, it might still do nothing to satisfy the CMEA's energy needs. It might even harm them.

As a general rule, whether a radical change in an OPEC country's orientation will be good for the Soviet Union depends on the country's economic circumstances. Thus, a revolution in the so-called 'poor' but heavily populated oil countries is likely to have less impact on the world economy once a new government takes over than a coup in a 'rich' but sparsely populated oil country. This holds whether or not a new government is more or less radical. Whatever direction a 'poor but populated' country like Iran, Algeria, Iraq, Indonesia, or Nigeria decides to go, the likelihood is that it will feel pressured to continue to import, to industrialize, and satisfy its own population's needs. To do this it will have to export. The only readily saleable commodity most of them have is petroleum. The new leaders may express considerable xenophobia and may demand, as has the religious leader Ayatollah Khomeini in Iran, to 'save our petroleum reserves for future generations'.[56] But unless the populace also agrees to give up the 'necessities' of the modern world not only in theory but in fact, any new government in this group of countries will have no choice after a time but to continue to export even though it be at a considerably reduced rate. Angola, which not only kept its contract with Gulf Oil but signed a new one with Texaco, illustrates the point.

In contrast, a revolution in countries rich in oil but with small populations like Saudi Arabia, Kuwait, and Libya would have an impact on the world energy market that inevitably affects the USSR. Countries such as Kuwait and Libya have already demonstrated they can cut back sharply on their oil output without imposing major sacrifices on their population.[57] If nothing else, they can use their accumulated earnings from past exports to live adequately on the interest and dividends.

To take an extreme but probably unlikely case, the radicalization of a country like Saudi Arabia in favor of the Soviet Union could be an important benefit to the Soviet Union. While the oil importers of the world suffered, the Soviet Union as an exporter would presumably benefit at least politically as the price of its petroleum products increased. Moreover, the Soviet Union's new ally might agree to divert an increased share of its reduced output to the Soviet Union. As in the case of Iraq, if nothing else this oil could be bartered as payment for the purchase of Soviet arms. In turn this would allow the Soviet Union to resell the oil as it saw fit.

The Soviets could benefit even if their new allies did not agree to supply the Soviet Union with resaleable oil. It would be damaging enough if the rest of the world came to believe that the Soviet Union had managed to dominate a formerly neutral or pro-Western oil producer. If the Soviets ever gain the power to co-ordinate a contemporary cut-back in production and export, that would be as effective a political weapon as the threatened use of atomic weapons or troops. Then it would not only be the Arab countries that would have the power to use oil to seek their political aims, but the Soviet Union. It is possible that the Soviets could turn Western Europe and Japan into political Finlands.

The above scenarios would certainly be welcomed by the Soviet Union, but there is no guarantee that it would always work out favorably for it. Indeed, there is good reason to believe that any change in the existing political make-up of the major OPEC countries is as likely as not to be adverse to Soviet interests. There is no assurance that all future revolutions will be limited solely to right-wing or neutral governments. It is just as possible that a country like Iraq, which has tended to be pro-Soviet, will be taken over by a government that is neutral or anti-Soviet. This does not necessarily mean that the West will suddenly be inundated with petroleum, but it could mean, for example, that if a government in a pro-Soviet country like Iraq were overthrown, the Soviet Union would no longer find it so easy to convert its weapons into hard currency by bartering them for Iraqi oil.

As the case of Iran suggests, even if a pro-Western government is overthrown or circumscribed, there is no guarantee that the resulting government will be pro-Soviet. Indeed, it may be neither pro-Soviet nor pro-Western if a religious Muslim government succeeds to power. Since most of the OPEC countries are Muslim, that is not a particularly propitious prospect for the Soviets. Presumably such a government, at least as long as it is poor and heavily populated, will, as indicated above, have to keep exporting at a level not too far below the pre-coup volume. The flow is likely to be more erratic for all parties and more subject to political disruption. Since some of the religious Muslim movements tend to be outspokenly anti-communist, the impact on the Soviet Union and Eastern Europe may be equally, if not more, disruptive than it is for the West. This does not follow automatically, as the coups in Iraq and Libya suggest, but in the case of Iran the Soviets have been at least as seriously affected by the chaos as the West. This is because the uprising curtailed exports not only of petroleum but of natural gas as well. As one of the Shah's opponents put it, 'the Shah says that his opposition is influenced by the communists, but he is the one who entertains communists and sells gas to the Russians at ridiculously low prices'.[58] Added another, 'Iran should not sell all its oil to the West and all its natural gas to the Soviet Union'.[59] While Western importers of petroleum were able eventually to shuffle their supply lines around, so that the flow of oil has remained relatively steady, the USSR, virtually the sole importer of Iranian natural gas, has had a more difficult problem. The soviet republics of Armenia and Georgia found themselves temporarily with a major source of their supplies suspended.[60] And as the Caucasian republics discovered, it is more difficult to arrange alternative sources of natural gas supplied by pipeline than it is to rearrange sources of more easily and flexibly shipped petroleum.

The Shah's overthrow also called into question the Soviet expenditures in connection with the IGAT II (Iranian gas trunkline). When the Iranians agreed to participate in this arrangement, it seemed to be a very rational decision. Most of the Iranian gas that was to be set aside for export could not be used in any other way. It was a by-product of their production of petroleum that either had to be consumed or flared and thus gone for ever. Since the Iranians could consume only a fraction of this gas at home it was logical to export it, and the only potential consumer in the neighborhood was the USSR. When the revolution came, petroleum production was cut back sharply and, consequently, so was the by-product natural gas. Since petroleum production did not equal the pre-revolutionary level, there was not enough by-product gas available to satisfy Soviet needs. Moreover, the Iranians announced that they planned to increase their own domestic consumption of natural gas. Thus they no longer needed the Soviet Union as a customer.

There seemed to be real doubt as to whether the Iranians would resume substantial shipments of their gas to the Soviet Union. Obviously this would jeopardize plans for IGAT II. This even raised some doubt as to whether the Soviets would ever again receive the full 9 billion cubic meters of gas they had been importing through IGAT I. Yet, there was some suspicion that this was all part of an elaborate bargaining strategy.[61] Even though in 1974 and 1977 the Shah had forced the Soviets to increase their payments for Iranian gas, the post-Shah government had expressed considerable displeasure that the Shah did not raise the price even more.[62] To some, it looked as though if and when the Soviets agree to pay a higher price not only for the gas in IGAT II, but in IGAT I, the Iranians would be more amenable.

This left the Soviets with a dilemma. They needed Iranian gas to supplement their supplies in the southern Caucasian republics. They were also counting on IGAT II to provide more gas not only for those republics but also for some of their industrial operations in the Donetz Basin.[63] Still, disruptions such as those that occurred in 1978 and 1979 created uncertainty about the wisdom of making the USSR dependent on imports from Iran. Moreover, the Soviet Union according to one report had completed its 500-kilometer portion of the pipeline in Iran that presumably was left to sit idle.[64] Also in question were the 1 million tons of petroleum that the Soviet Union was to receive as payment for the construction of that pipeline.

In the aftermath of this misadventure there must be some Soviets who question the whole arrangement with Iran. Why did the Soviets have to jeopardize their traditional policy of self-sufficiency? Certainly, since Soviet gas reserves are so vast, they could easily divert some of their own gas to the Donetz and the Caucasian republics. At the same time, there is reason to believe that whether or not the Iranians decide to participate in IGAT II, the Soviets are obligated to fulfill their part of the deal that would involve exports of at least 15 billion cubic meters of gas exports to Czechoslovakia, Germany, Austria, and France.[65] We know the Soviets had the pipeline capacity built into the

Orenburg–Soiuz pipeline. They probably also have the surplus gas, although initially it might be hard for them, since they probably would have had to divert some of their gas shipments to the south as well. But if they could handle both their own needs and the exports, why should they have bothered with Iran? For some, that was undoubtedly an embarrassing question.

WHAT SHOULD US POLICY BE?

While the Soviet experiment in Iran suggests that the Soviet Union cannot necessarily count on the Middle Eastern countries to supplement its energy supplies, neither does it necessarily mean that the Soviet Union will never be able to import large quantities of energy from the Middle East. Certainly, if the Soviets do become regular and major customers in the Middle East, this will have important implications for the United States. Inevitably their purchases will reduce the size of petroleum supplies.

While it might be comforting to see that the Soviet Union is no longer so self-sufficient, that increased dependency can only come at someone else's expense, and it is all but inevitable, therefore, that this change in relationships will cause increased economic and political strain elsewhere. In the circumstances, a major question is whether the United States should facilitate or frustrate Soviet energy development. From what we have seen, there is no doubt that, if the Soviets are to realize the full potential of their energy resources, they will have to avail themselves of foreign technology and help. Under the circumstances, should we support, ignore, or block the transfer of that technology?

One influential group of critics argues that the United States should closely regulate and normally ban the sale of all technology, and energy technology in particular, to the Soviet Union. These arguments have found their explicit expression in the Amendment attached to the 1974 Stevenson Amendment to the Export-Import Bank Act of 1945. These Amendments restrict the money that might be lent by the United States Export-Import Bank to finance 'the purchase, lease, or procurement of any product or service which involves research and exploration of fossil fuel energy resources' in the Soviet Union to $40 million.[66] For that matter, there can be no 'loan or financial guarantees ... for the purchase, lease, or procurement of any product or service for the production (including processing and distribution) of fossil fuel energy resources'. Thus, the Export-Import Bank could lend small sums of money for research and exploration but not for production.[67]

There are those who argue that not only should American government agencies frustrate the sale of such technology, but they should prohibit such sales outright. Such critics, like Carl Gershman, claim that the sale of the drill bit manufacturing plant by Dresser Industries of Texas in 1978 for $144 million not only will make it possible for the Soviet Union to increase its petroleum production and thereby its economic strength, but will also increase Soviet military potential. One of the components of the drill bit process involves producing tungsten carbide which, according to Gershman, 'could produce armor-piercing projectiles'.[68] Gershman also supports the charge that with the drill bit factory the Soviets will be able to increase their export of drill bits to the Middle East and thereby expand their political influence there. Thus, in one transaction, the United States has enhanced the Soviet Union's economic, political, and military capability. Given that a task force of the Defense Science Board concluded 'the deep well technology in question was wholly concentrated in the United States,' Gershman argues that, if we had refused to sell, the Soviet Union would have had nowhere else to turn. Thus we could have continued to sell them individual drill bits, but only when it suited us.

More moderate spokesmen who want to restrict the transfer of petroleum technology would permit the sale now and then of such technology, but only where it can be used to increase our economic and political leverage.[69] This will be effective, as Samuel Huntington argues, because the Soviets engage in economic and political trade-offs when they feel the benefits are worth the cost. That is not to say that even this somewhat more moderate view of strategic control is necessarily easy to administer. Huntington, like Gershman, was also opposed to the sale of the Dresser drill bit plant.

Yet it turns out the export license for drill bits was used by the United States as a political weapon. Although not widely publicized, permission to sell the drill bit plant apparently was linked to the release from a Moscow jail sentence of F. Jay Crawford, a business representative of International Harvester. Crawford was arrested during the touchy summer of 1978. His arrest on 12 June was followed a few weeks later by the sentencing of several Soviet dissidents, including Anatoly Shcharansky and Alexander Ginsburg. The combination of several dissident trials and the arrest of Crawford led President Carter on 18 July 1978 to revoke an already issued license for the export of a Sperry-Univac computer ordered by TASS for use during the Moscow Olympics. At the same time, Carter reinstituted a licensing requirement for the sale of gas and oil technology to the Soviet Union. As a result of this new regulation, the Dresser contract for the sale of the drill bit plant became a pawn in the American government's show of outrage over Soviet actions.

The business community was bitterly distressed that it had been caught up in this political battle. In its eyes, it would inevitably cause it to lose important sales. None the less, the American government's position was that it was unseemly and politically indefensible to conduct business as usual with an American businessman in jail and the Soviets flaunting all international codes of treatment for their dissidents.

Caught in the middle, the whole episode for Dresser was a bewildering one. After all the drill bit plant had been under discussion with the Soviets since 1973. Thereafter, negotiations were slow, but persistent. First the Soviets decided to seek comparable bids from companies in the United States as well as one in France. Yet, once the bids were in, no decision was made, because it had become clear that, while the Ministry of the Petroleum Industry was most eager to buy the plant, Gosbank, Gosplan, and the Ministry of Foreign Trade refused to allocate enough hard currency for the purchase. Only in late 1977 did Gosbank relent and authorize the release of the funds that made possible the signing of the $144 million preliminary agreement in November 1977. A more formal second agreement was signed in March 1978. With this in hand, Dresser then sought an export license. A license for everything but the computerized electron-beam welding component was issued in late May. On that basis the final contract was signed a few days later. The arrest of Crawford followed less than two weeks later and the Shchransky trial shortly after that.[70]

When Carter announced his decision to curb sales to the Soviet Union on 18 July, it was unclear what exactly would happen to the Dresser contract. Dresser finally got what looked to be permission to export on 9 August 1978, when its export license for the electron-beam welder was approved. But opponents of the sale in the National Security Council, the Department of Energy, the Department of Defense, the United States Senate, and American industry began to protest. As they saw it, the sale increased Soviet military strength. It also provided the Soviet Union with a computerized electron-beam welder, an extremely sophisticated piece of equipment.

Although the license was not revoked, the whole matter was sent back to a senior inter-agency committee for examination on 28 August. But the committee labored in vain. While it was reflecting, the Soviets, after several high-level messages from President Carter, suddenly moved Crawford's case to trial on 5 September. (Ironically the Soviet court agreed, although grudgingly, to postpone the trial for one day, so the Americans could celebrate Labor Day.) Word had already been sent to the United States that the court would find Crawford guilty and then expel him. The next day, on 6 September, Carter authorized the export of the plant. Permission was publicly announced on 7 September, and the following day as promised, having been found guilty, Crawford was expelled from the Soviet Union, a free man.

Much of the story remains to be told. Why was it that Crawford was arrested? What was the connection between his case and what appeared to be the contrived arrest of three Soviet spies in the United States? How did the United States business community react to Crawford's arrest? What explains the Soviet prosecutor's and the KGB's embarrassing lack of evidence? But that is another tale. For now it is enough to know that the Dresser contract was used as a lever to influence Soviet behavior. Huntington and some of the others probably would have been happier if the drill bit plant had been used for broader political purposes in addition to helping facilitate the release of Crawford. Yet there are many others who insist that such leverage can only be pushed so far and, unless the pushers are cautious, there is a danger that not only will the leverage be lost, but the sale as well.

Those who worry that the United States may be the ultimate loser if petroleum technology is carelessly used as a political lever against the Soviet Union focus on several issues. Most of all they argue that those who want to use trade as a lever suffer from an informational lag. The levergers do not seem to understand that the United States no longer dominates world technology as it did in the 1940s and 1950s when, if the United States decided to embargo a product, there was virtually nowhere else to turn. Today there are numerous alternatives even in petroleum technology. The United States still leads the field of petroleum technology, but increasingly that domination is being challenged. Today only a very small portion of the petroleum equipment market is monopolized by Americans. Estimates of how large that fraction of the market is vary, but almost all who have considered the matter say the United States has exclusive control of no more than 15 per cent of the market. Moreover, there is virtually nothing the United States has that cannot be replaced with a substitute from a manufacturer in some other country. That substitute may be second best, but it will do. One of the few items that is produced only in the United States is the blow-out preventer. That is not a critical piece of equipment. If we banned its export, the Soviet Union would just have to contend with more blow-outs. In other words, much of that uniqueness refers to items that are useful but not essential. The main advantage in buying American equipment is its reputation, its price, the experience which has gone into its making, and the scale of our production which most closely approximates what the Soviets are trying to do. These are important considerations but not enough to deter the Soviet Union from buying elsewhere if forced to.

Indeed, if anything there is some critical equipment that is not manufactured in the United States. For example, the United States must rely on Schlumberger of France for logging equipment. Given the spread of technology around the world, the United States may find that withholding an export license will simply result in the loss of a sale. One of the reasons for Dresser's furor over the withholding of the license was that in its eyes, the levergers' threat to withdraw its export license was predicated on what Dresser took to be unfamiliarity with the market. Contradicting the claims of the United States Department of Defense, Dresser asserted that the Soviets and some other East Europeans already knew how to manufacture tungsten carbide. As for the electron-beam welder, it had actually been invented by the French and prototypes had been long ago purchased by the Soviets (although without the computer attachment).[71] For that matter Dresser was concerned that the whole contract would be grabbed up by the French. After all, a French company had also been asked to bid on a bit drill plant back in 1974.

The loss of some important sales is more than an idle concern. We saw earlier that the Soviets had placed a high priority on purchasing gas lift equipment to improve their secondary recovery efforts in west Siberia. They first turned to one of the industry's leaders, NL Industries of Texas. However, under Carter's 17 July 1978 ruling, the export of petroleum equipment required a license. NL Industries dutifully applied for the appropriate permission. An agreement to buy from NL Industries was reached, but after a time the Soviets told the Americans that unless the license was approved in two weeks' time, they would order elsewhere. As it turned out, the US license was issued four weeks later, but the Soviets had no way of knowing and decided not to wait. As they had threatened, after two weeks they cancelled their $226 million order with NL Industries and, in late October 1978, bought their gas lift equipment from Technip, a French company, instead.[72] The sale of a sonar device was similarly lost when the Soviets turned to a Norwegian firm.

To demonstrate the wide availability of petroleum technology, Armco has drawn up a list of parts that went into the construction of the deep water semi-submersible drilling rig it sold to the Soviet Union. In a report prepared for the Sub-Committee on International Economic Policy and Trade of the Committee for Foreign Affairs of the United States House of Representatives, Armco found that all 2,500 major component parts as well as the millions of individual pieces of equipment that comprised the drilling rig were available outside the United States.[73] American companies are not the only ones capable of drilling offshore. The Soviets have already authorized offshore drilling in Soviet waters by Japanese, French, and British companies. They are talking to others as well.[74]

Finally, for those who still have doubts, the CIA has tabulated which countries have sold the Soviets petroleum and gas equipment. If large-diameter pipe is excluded (which the Soviet Union normally does not buy from the United States), the United States from 1972 to 1976 sold

$550 million worth of equipment.[75] However, that amounted to only 18 per cent of total petroleum and gas equipment sales (exclusive of $4 billion worth of large-diameter pipe) from OECD countries to the USSR during that period. True, some of the equipment was from foreign-based subsidiaries of American companies, but today our allies are increasingly less tolerant of US attempts to restrict export opportunities of overseas American subsidiaries. In the case of the French, the government has even gone so far as to take over one reluctant American affiliate.[76]

IS IT IN THE UNITED STATES' INTEREST TO INCREASE SOVIET OIL AND GAS PRODUCTION?

Even if the United States could control the export of technology to the Soviet Union, would it be in our interest to try to do so? Would we want to exercise such power so rigorously that we prevented the Soviets from increasing their petroleum and gas production? Gershman and others argue that anything that holds the Soviet growth back by even a few percentage points is in our national interest.[77] After the Soviet invasion of Afghanistan, they presumably feel even more convinced of the need for such a strategy, and certainly an embargo on US technology sales would be a clear sign of US displeasure over Soviet aggression. But would this be such a wise policy? Certainly there is a danger in building up the Soviet economy, but there is also an advantage in increasing Soviet oil production, particularly when the Soviets have consistently exported over 25 per cent of it. Anything that can be done anywhere in the world to increase petroleum production would seem to be in our long-range interest. The more petroleum there is available for sale and the more sellers there are, the less vulnerable we become to economic blackmail. The more producers there are, the more likely it is that some of them will go their own way unfettered by the dictates of a small group. Because the Soviet Union is our main ideological and power rival in the world today, is that reason to curb its output? Is it any less reliable as a supplier than Iraq or Libya? For that matter, even faithful Nigeria confiscated British Petroleum's interest in some Nigerian fields, because the United Kingdom was exporting to South Africa against Nigerian wishes.

Or go to the other extreme: assume that the Soviet Union finds that instead of being able to export 1 million barrels a day to the hard-currency world and about half a million barrels a day to the non-communist developing countries as it has been doing, it is forced to go out into the market and by 1985 import 2·7 to 4·5 MBD. If the Soviets are successful in finding that much petroleum in the market, it would mean that the non-communist world would find itself fighting to fill its tanks with 4·2 to 6 MBD less petroleum in the pipeline (1·5 MBD less in exports and 2·7 to 4·5 MBD more in Soviet imports). The gasoline panic after the Iranian revolution was set off by the disappearance of less than that. On balance, an equally good argument can be made for helping the Soviets sustain their status as a world petroleum exporter. Finally, since we do not have the control over technology that we once had, we may not have much choice in the matter.

SOVIET ISOLATIONISM

The choice may not be entirely ours for other reasons. Invariably in such matters we view such options from our myopic vantage point. It is an effort for us even to concede that our OECD allies have not only the capability but the desire to act out of concert with American policy. Not surprisingly, therefore, almost no one has suggested that some Soviets may also have some independent thoughts about the matter. We have assumed all along that the Soviets will continue to export petroleum and gas on a regular basis to pay for their imports. Yet, the fact that the Soviet Union now finds itself increasingly dependent on foreigners is regarded with considerable distaste by some in the USSR. Such sentiments are not necessarily limited to a few officials. Resentment is particularly widespread when, because of the need to export, domestic consumption is affected. The sharp increase in the export of petroleum to reduce the trade deficit is the best example of how the export market has come before domestic needs. As noted earlier, the sharp jump in exports occurred in 1975, when shipments rose by 12 per cent (18 per cent to OECD countries) and in 1976, when they increased by 15 per cent (29 per cent to the OECD). The effect on the growth of domestic consumption was perceptible. Whereas until 1974 domestic consumption of petroleum never increased less than 7 per cent a year, in 1975 it only increased by 5·8 per cent. That was less than the 7 per cent increase in production but was still large enough to be accommodated without too much difficulty. However, in 1976 domestic consumption increased by only 2·9 per cent and in 1977, by 3·6 to 4·7 per cent. In both instances there were reports of shortages of gasoline that seemed to transcend the usual complaints of inept planning procedures. Not only can such diversions inconvenience those fortunate to have an automobile in the Soviet Union, but it can also have a direct impact on economic growth. Undoubtedly this shortage of petroleum has contributed, at least in part, to the recent decrease in Soviet economic growth rates.

Reacting to a variety of such developments, some in the Soviet Union have warned about the danger of too much interchange with the West.

Some even go so far as to seek a cessation of most, if not all, trading relationships. As they see it, any step toward increased Soviet economic interdependence with the capitalist world is ultimately bound to be economically disadvantageous and also may be disruptive to the traditional Russian cultural and social structure. The diversion of Soviet petroleum from domestic to foreign markets does nothing to allay the first set of such fears. This exploitation, some call it rape, of Russian natural wealth for the benefit of Western multinational corporations is what these opponents of increased trade have campaigned against. They claim that the benefits of increased entanglement with the West are ephemeral. The technology purchased becomes outdated quickly while the Russian raw materials, wastefully squandered by aliens in the West, are lost for ever to future Russian generations.

For those familiar with Russian history, much of the present debate will seem like a continuation of the old argument between the Slavophiles and the Westernizers. The Slavophiles of the nineteenth century urged Russia to turn its back on the West. Failure to do so, they argued, would open Russia's borders not only to Western goods but Western ideas and ways of doing things. That would mean slums and strikes as well as degradation, disruption, and ultimately social unrest. (Since Marxism was a product of the West, the Slavophiles may not have been entirely misguided.) Instead Russia, with its great population and natural wealth, would be better advised to follow its own path of development. From their point of view, Russia should adhere to the traditional Russian way of doing things, looking for guidance to such indigenous institutions as the Russian peasant and the Russian church. Russia should evolve in its own way with its own timing.

The modern-day version of the debate is most eloquently reflected in exchanges between Alexander Solzhenitsyn and Andrei Sakharov. Obviously neither one can be considered an official spokesman for anything in the Soviet Union, but their views none the less find support throughout the Soviet system. Solzhenitzyn, in a letter dated 5 September 1973, resurrected the banner of the Slavophiles. He urged Soviet leaders to turn their backs on the outside world and concentrate on internal Soviet development. He called for an end to the stress on rapid industrialization and urged a halt to further sales to the West of Russia's natural resources, such as Siberian natural gas, oil, and timber. As he stated: 'We, a great industrial superpower, behave like the most backward country, by inviting foreigners to dig our earth and then offer them in exchange our priceless treasure – Siberian natural gas.' In fact, he wants a 'Russia first' policy of 'let's save our raw material patrimony for future Russian generations'. That same attitude is reflected by Boris Komarov in his *samizdat* book *The Destruction of Nature*. The raw materials will always be valuable, but the Western technology will soon become obsolete. Why give up something timeless and valuable for something of a temporal value?

In response, Sakharov argued that such a policy would be isolationist. As Sakharov put it: 'Our country cannot exist in economic and scientific isolation without world trade, including trade in the country's natural resources or divorced from the world scientific technical progress – a condition that holds not only danger, but at the same time the only real chance of saving mankind.'[78]

Besides unofficial spokesmen like Sakharov and Solzhenitzyn, there are others expressing the same clash of opinions who represent a more official point of view. One extreme case is the *Pravda* article written by Professor K. Suvorov, who is believed to serve as a consultant to the Central Committee of the Soviet Communist Party. In his plea for Soviet economic independence, he seemed to go beyond urging economic autarky for the CMEA, to calling for a return to Stalin's version of socialism in one country.[79] He even cited Stalin as the originator of such an idea. The reference to Stalin, which originally was printed in *Pravda*, was thoughtfully omitted in an otherwise fairly complete translation of the article in *Soviet News* (the news bulletin of the Soviet Embassy in London).[80] In Suvorov's view, Stalin wanted the Soviet Union 'to steer the course towards the country's industrialization, the development of production of the means of production, and the formation of reserves for economic maneuvering' so as to ensure the Soviet Union's economic independence from the world's capitalist economy and achieve the complete triumph of socialism. This 'industrialization of the USSR would ensure the economic independence of the country and the ousting of capitalist elements from all the sectors of the national economy, consolidate the Soviet Union's economic and defense potential, and strengthen friendship among the peoples.'

Without taking such an extreme stand, others worry that the Soviet Union may be overexploiting its natural resources and wealth for the benefit of foreigners and at the expense of future Russians. The emphasis on future generations has also become a recurrent theme of politicians and economists.[81] For example, American sponsors of the Neftegas (oil and gas) exhibition of American petroleum and gas equipment held in Moscow in October 1977 were unprepared for some of the questions asked by reporters at a press conference called specifically to promote the sale of such equipment. What does one say when a Soviet questioner asks, 'Won't the use of this equipment lead to the depletion of our Soviet resources?' The American conference organizer assumed he had misunderstood the question and asked that it be repeated. Even those who accept the need to exploit Soviet raw materials because they want Western technology warn that such a policy is not as simple as it seems, and that it necessitates ever-increasing expenses because of the need to go farther to the north and the east.[82]

There is even some reason to believe that the debate extends into the Politbureau itself. Of course, there is a danger in placing too much emphasis on the slightly different utterances made one day by Party Secretary Brezhnev and those made a few days later by Prime Minister Kosygin. None the less, in October 1974 Brezhnev gave a speech in which he said, 'The natural resources of our country allow us to look to the future without danger. To make a long story short, our country is a country with uncounted riches and *inexhaustible* opportunities. It is our job to use these riches and opportunities properly and economically.'[83]

Three weeks later, Kosygin seemed to view the situation in a different perspective. 'Our country is provided with everything necessary, so that the Soviet economy can develop dynamically ... Our resources are great. But they are *not inexhaustible*. They belong not only to the present but to the future generation of Soviet people. Therefore it is our task to use them intelligently, carefully, and in the most rational way possible, so that each kilogram of fuel, metal, cement, cotton, fertilizer, synthetic material ... all of these serve the Socialist economy as effectively as the most advanced raw material technology permits.'[84]

Shortly thereafter, Brezhnev apparently changed his attitude. He too began to call for more care in the exploitation of raw materials. 'The demand of the country for energy and raw materials grows increasingly and, therefore, production becomes all the more costly. Consequently, if we are to avoid an extraordinary increase in capital investment, it is necessary to use raw materials more effectively.'[85]

Of course, it is never clear if the leaders of the Soviet Union ever pay attention to or even care what each other say or read. But if they do, or if they read *Pravda*, they might notice the sharp difference in attitudes toward the use of raw materials.

Finally, those who justify the continuation of raw material exports sometimes adopt a novel rationalization. A. A. Trofimuk, deputy director of the Siberian section of the Soviet Academy of Sciences, urged even faster exploitation of oil and gas.[86] Concerned that it is only a matter of time before new energy substitutes are found, he proposed that the Soviet Union use its reserves now before they become valueless.

CONCLUSION

While Soviet conservation efforts might be facilitated if Soviet enterprise managers responded more to price changes and other market mechanisms, the Soviets have other methods of ensuring that, if need be, petroleum and gas can be set aside for high-priority uses including export. The very absence of a meaningful price change has meant that the Soviet Union and Eastern Europe have only recently begun to reduce their use of coal. As a result, CMEA bloc economies have not yet had a chance to burn their public transportation bridges behind them. Almost all of the public transport and coal infrastructure is still in place.

Because Soviet individuals consume relatively little energy in cars, some outside observers believe that conservation will not come easily. Since the USSR has never enjoyed the luxury of being very wasteful, there is not much the Soviet or East European consumer can do to improve its ways. The economizing, if it is to come, must come from industry. In contrast to the United States, where industrial usage accounts for less than 35 per cent of all energy consumption, industrial consumption is closer to 60 per cent in the Soviet Union. In a sense this is good news, because it means the target is concentrated, and that there are fewer decision-makers to influence. At the same time, it is bad news, because these decision-makers are locked into a planning system that rewards increasing production rather than decreasing costs. Yet the Soviets are making some progress, as evidenced by the fact that in 1977 energy consumption per unit of GNP levelled off and began to fall.

The Soviets also have an ample supply of substitutes for their petroleum. True, the bulk of these reserves are in the far eastern and northern parts of the Soviet Union, but since they have had more experience working in a cold climate, they seem better able to cope with that than we do. Moreover, the ever-climbing price of energy works its way through the increasingly ineffective insulation of the non-convertible ruble. Finally, even if all else fails, the Soviets are determined to move ahead with nuclear energy.

We may ridicule the Soviets for many things, but unlike some of their rivals in the West the Soviets do have an energy policy. We might have one too, if there were neither an environmental lobby to protest about the nuclear radioactivity or the coal sulphur emission hazards nor a consumer lobby to protest that higher prices will unfairly result in windfall profits for producers and exporters. Regardless of whether or not we think the Soviet system is inhumane or callous, the point is that the Soviets can, if need be, move decisively.

Much more uncertain is what our policy should be. We not only have trouble deciding what our energy policy should be inside the United States, we also cannot decide whether we should facilitate or obstruct the development of Soviet energy. For that matter, how much longer will the Soviets continue to seek our technology? Normally they come to us first because of our longer experience, and because the scale of our operations more closely approximates their efforts. This gives us some bargaining power. In the circumstances there is no doubt that, when

Soviet behavior is particularly objectionable, as in Afghanistan, we should attempt to use the sale of that equipment as a lever to influence Soviet action. Sometimes we must do that to satisfy our own sense of decency even if we know in advance that our efforts may in the long run become counter-productive, and even though we know that as technology spreads around the world our leverage is bound to be less effective.

Generally, however, if leverage is to have an impact, it must be used infrequently and judiciously. That is not only because the Soviets realize they can fall back more and more on second-best but adequate substitutes. It is also because it seems to be in our long-run interest to facilitate Soviet interest in energy production. If we do not supply them, somebody else will. Furthermore, the more energy they produce, the more they are likely to sell in the world market and the longer it will take before the Eastern Europeans and the Soviets start competing with us in the world market for petroleum.

NOTES: CHAPTER 7

1 Department of Energy, *Monthly Energy Review*, June 1979, p. 12; Robert Campbell, *Soviet Energy Balances* (Santa Monica: Rand Corporation, 1978), p. 22; *Sotsialisticheskaia industriia*, August 8, 1979, p. 2.
2 Campbell, 1978, p. 27.
3 Nekrasov and Pervukhin, p. 88.
4 *Petroleum Economist*, September 1978, p. 363.
5 *Pravda*, 14 June 1978, p. 2, 10 November 1979, p. 2.
6 Robert J. McIntyre and James R. Thornton, 'Urban design and energy utilization: a comparative analysis of Soviet practice', *Journal of Comparative Economics*, no. 4, December 1978, p. 351; *Pravda*, 10 November 1979, p. 2.
7 *Tiesa*, 13 May 1977, p. 2; *Current Digest of the Soviet Press*, 1 June 1977, p. 23, 19 October 1977, p. 21; *Pravda*, 10 November 1979, p. 2.
8 *Selskaia zhizn*, 28 February 1978, p. 3; *Izvestiia*, 10 July 1979, p. 3.
9 CIA, August 1978, p. 6.
10 A. Troitskii, 'Elektroenergetika problemy i perspektivy', *Planovoe khoziaistvo*, February 1979, p. 25.
11 N. Tikhonov, 'Sovershenstovanie upravleniia-vazhnyi reserv povysheniia effektivnosti obshchestvennogo proizvodstva', *Kommunist*, no. 7, May 1979, p. 44.
12 Lalaiants, p. 6.
13 L. Iofina, 'Prognozirovanie potrebnosti v energeticheskikh resursakh v ASPR Gosplana SSSR', *Planovoe khoziaistvo*, November 1977, p. 133; *Sotsialisticheskaia industriia*, 8 August 1979, p. 2.
14 *Sotsialisticheskaia industriia*, 13 October 1978, p. 3; Probst, p. 73; A. M. Nekrasov, 'Osnovnye napravleniia ratsional'nogo ispol'zovaniia i ekonomii energoresursov', *Planovoe khoziaistvo*, May 1978, p. 76.
15 *Pravda*, 3 July 1979, p. 3.
16 *New York Times*, 12 August 1979, s. IV, p. 6.
17 ibid., 7 January 1975, p. 1.
18 Leslie Dienes, 'The Soviet union: an energy crunch ahead?', *Problems of*

Communism, September–December 1977, p. 43; John P. Hardt, Ronda A. Bresnick and David Levine, 'Soviet oil and gas in the global perspective', *Project Interdependence: US and World Energy Outlook through 1990* (Washington, DC: Congressional Research Library of Congress, November 1977), p. 803.

19 *Ekonomicheskaia gazeta*, no. 3, January 1979, p. 2; *Current Digest of the Soviet Press*, 7 February 1979, p. 21; *Ekonomicheskaia gazeta*, no. 34, August 1978, p. 2.

20 *Pravda*, 16 January 1978, p. 2; *Izvestiia*, 4 December 1978, p. 1.

21 Stern, p. 39; Martsinkevich, p. 64; *Petroleum Economist*, August 1979, p. 313.

22 *Ekonomicheskaia gazeta*, no. 24, June 1978, p. 2; *Gas and Oil Journal*, 22 May 1978, p. 33; Jensen, pp. 23–32.

23 *Petroleum Economist*, April 1979, p. 163.

24 Stern, p. 39; CIA Gas, July 1978, p. 6; *Review of Sino-Soviet Oil*, November 1978, p. 27; *Petroleum Economist*, August 1979, p. 314.

25 Donald L. Mueller, 'Siberian natural gas for the US', paper delivered at the Eastern AAASS Meeting in Storrs, Connecticut, 1 May 1976, p. 13.

26 *Oil and Gas Journal*, 25 December 1972; *Review of Sino-Soviet Oil*, April 1979, p. 3; *Foreign Trade*, August 1976, p. 15.

27 Dienes and Shabad, p. 255; *Izvestiia*, 9 August 1979, p. 13; *Pravda*, 19 August 1979, p. 2.

28 Boris Komarov, *Unichtozhenie prirody* (Frankfurt; Posev, 1978), p. 156.

29 American Petroleum Institute, *Petroleum in the Soviet Union* (Washington, DC: API, 1972), p. 51.

30 *Izvestiia*, 24 February 1978, p. 2; *Sotsialisticheskaia industriia*, 10 August 1979; p. 2.

31 *Pravda*, 16 January 1978, p. 2.

32 ibid., 14 June 1978, p. 2.

33 CIA Handbook 1978, p. 84.

34 Dienes and Shabad, pp. 105–6; Hardt, Bresnick and Levine, p. 801.

35 Martsenkevich, p. 64.

36 Mazover, 1977, p. 141.

37 *Current Digest of the Soviet Press*, 20 June 1979, p. 17.

38 Philip Pryde, in Dienes and Shabad, pp. 151 and 156; *Current Digest of the Soviet Press*, 22 June 1977, p. 21.

39 Campbell, *Soviet Energy Balance*, p. 14.

40 Nekrasov and Pervukhin, p. 114.

41 ibid., p. 114; CIA, August 1978, p. 6; *Time*, 30 October 1977, p. 69.

42 *Izvestiia*, 1 February 1979, p. 1; *Current Digest of the Soviet Press*, 3 January 1979, p. 17; *Petroleum Economist*, June 1979, p. 254; August 1979, p. 340; *Economicheskaia gazeta*, no. 2, January 1980, p. 1.

43 Pryde, p. 161; Novikov, p. 33; *New York Times*, 14 January 1980, p. A6.

44 Pryde, p. 164.

45 Robert W. Campbell, *Soviet Energy R&D: Goals, Planning, and Organizations*, R-2253-DOE (Santa Monica: RAND, May 1978); *Izvestiia*, 11 April 1979, p. 2; *Pravda*, 15 July 1979, pp. 27–32.

46 *Izvestiia*, 11 April 1979, pp. 2 and 3.

47 Nekrasov, p. 78.

48 Pryde, p. 158; *International Herald Tribune*, 18 October 1978, p. 4; *Soviet Weekly*, 7 July 1979, p. 5.

49 Campbell, *Soviet Energy R&D*, pp. 32–6; see also *Pravda*, 7 August 1979, p. 3; *New York Times*, 14 January 1980, p. A6.

50 *Soviet News*, 15 May 1979, p. 151.

51 *New York Times*, 23 April 1979, p. A15, 26 August 1979, book review, p. 11; *Washington Post*, 15 July 1979, p. H1.

52   *Izvestiia*, 11 April 1979, pp. 2 and 3.
53   Campbell, *Soviet Energy R&D*, p. 21.
54   Report of American businessmen.
55   *Petroleum Economist*, April 1975, p. 13; Campbell, *Soviet Energy R&D*,
      pp. 53-4.
56   *New York Times*, 23 November 1978, p. 2.
57   ibid., 7 January 1979, p. 10.
58   ibid., 9 December 1978, p. 3.
59   ibid., 19 November 1978, p. 1.
60   ibid., 20 November 1978, p. 84, 23 November 1978, p. 2; *Izvestiia*, 2
      December 1978, p. 5.
61   *Business Week*, 20 August 1979, p. 40.
62   *Wall Street Journal*, 30 July 1979, p. 23; Radio Liberty Research, 2 March
      1979.
63   *Soviet Export*, vol. 3, no. 120, 1979, p. 54.
64   ibid., vol. 5, no. 116, 1978, p. 21.
65   Jonathan P. Stern, *Soviet Natural Gas in the World Economy*, Washington, DC,
      Association of American Geographers, no. 11, 9 June 1979, p. 43.
66   Export-Import Bank Act 1945, Washington, DC, the Export-Import Bank,
      p. 19.
67   ibid.
68   Carl Gershman, 'Selling them the rope: business and the Soviets', *Commentary*,
      April 1979, p. 43.
69   Samuel P. Huntington, 'Trade technology and leverage: economic diplomacy',
      Foreign Policy, no. 32, Fall 1978, pp. 72-3.
70   *New York Times*, 2 June 1978, p. D5.
71   *Washington Post*, 8 September 1978, p. 86.
72   'Energy in the Soviet Union, a second look', conference held by the Russian
      Research Center, Harvard University, 6 November 1978; *International Herald
      Tribune*, 6 October 1978, p. 9.
73   'List of foreign source drilling equipment for semi-submersible exploratory
      drilling vessel capable of drilling in six hundred feet of water to a depth of 25,000
      feet', Washington, DC, Armco, 22 February 1979.
74   *Review of Sino-Soviet Oil*, July 1978, pp. 38-43.
75   CIA, July 1977, pp. 27-8.
76   *Business Week*, 17 May 1976, p. 47; *Wall Street Journal*, 5 November 1976,
      p. 6; *Moscow Narodny Bank Bulletin*, 14 March 1979, p. 14.
77   Gershman, p. 43.
78   *New York Times*, 15 April 1974, p. 1; *New York Review of Books*, 13 January
      1974, pp. 3-4.
79   *Pravda*, 18 December 1975, p. 2.
80   *Soviet News*, 13 January 1976, p. 15.
81   *Soviet News*, 15 July 1975, p. 242; Iakovetz, 1978, p. 77.
82   Iu. Iakovetz, 'Dvizhenie tsen mineral'nogo syr'ia', *Voprosy ekonomiki*, June
      1975, p. 3.
83   *Pravda*, 12 October 1974, p. 2, emphasis added.
84   *Pravda*, 3 November 1974, p. 2, emphasis added.
85   *Sotsialisticheskia industriia*, 24 March 1978, p. 1.
86   Dienes, 1977, pp. 57-8.

8

Shortcomings, Solutions, and the CIA

Analysis of future Soviet energy prospects involves more than a simple determination of whether Soviet petroleum deposits are either empty or full. Soviet energy planners are faced with serious problems but that does not necessarily mean they are unsolvable. In addition there are many unknowns that eventually could make an enormous difference. With a little bit of luck and a considerable amount of reorganization, the Soviet Union may remain a major petroleum exporting country. Yet with some bad luck and no change in the bureaucratic structure, the 1977 CIA report could prove to be correct and the Soviets, along with their East European dependants, may find themselves becoming net importers of large quantities of petroleum. While it is impossible to predict exactly what will happen in the years ahead, there is reason to believe that the Soviets will indeed be able to solve at least some of their problems. Moreover these solutions will be forthcoming sooner than might otherwise have been the case because of the unanticipated impact of the April 1977 CIA report. Therefore in this final chapter we will examine why the Soviets seemed to respond so slowly to their energy problems, what caused the CIA to issue the kind of report it did when it did and what the effect of the report has been.

PROBLEMS

The problems, as painstakingly described by the CIA, are real. Most of the petroleum fields in the older producing areas of the country are yielding less than they once did. Largely through overpumping, the recovery rate has in several instances decreased sharply. Secondary recovery measures, while sensibly used in many of the Volga-Ural well sites, have been senselessly applied elsewhere. Because of the pressure to maximize short-run output, the long-run maximum efficiency rate of recovery (MER) has often been sacrificed. As one CIA observer put it, the Soviets were producing petroleum in 1979 at a rate that under a MER set of guidelines would have been better produced in 1985. As we saw, to compensate for the production shortfalls elsewhere the

Soviets have been increasing output at Samotlor. By 1979 Samotlor provided about one-quarter of total Soviet output. If output at Samotlor should also fall, overall Soviet output will fall off sharply. Yet because of the counter-productive effect of water injection, that is exactly what the CIA and some Soviet authorities expect will happen.[1] Since there is no other back-up field with such a reserve potential, total Soviet output in the mid-1980s will decline by as much as 17 per cent to 33 per cent of what it was in the late 1970s. The consequence of losing Samotlor would not be so serious if the Soviets had moved vigorously in the mid-1970s to find new producing fields. Instead they neglected exploratory drilling and concentrated on development and supplementary drilling in already known fields. Again, that solves a short-run need but does nothing for long-run supplies.

THE FETTER OF MARXIST IDEOLOGY

As we have seen, much of the difficulty the Soviets have had in providing for their long-run petroleum needs is due to the irrational effect of Marxist ideology. It is worth summarizing this impact to gain an appreciation of just how pervasive it is. That Marxist ideology has caused so many problems is ironic because Marxists insist that it is capitalism, not communism, that leads to petroleum production at more than the MER. As they see it, capitalism has no built-in mechanism for coping with 'common' problems. Capitalism forces each private owner to pump as much as he can from a common field or reservoir for fear that if he does not his neighbors will beat him to it. Unfortunately, excessive drilling set off by competition of this sort causes a loss of underground pressure which reduces the ultimate quantity of petroleum that can be extracted.

Presumably a communist state with no private ownership should be able to prevent such unco-ordinated drilling and for the most part it has. Yet while avoiding one problem, communist societies somehow seem to come up with new and equally bothersome sets of problems. As one disillusioned party member put it, 'The advantage of communism is that it is always seeking to solve the problems that other economies never have.' True enough, the Soviets do not have to worry about co-ordinating the efforts of different property-owners. Yet they do have to worry about restraining their pumping crews so they will not exceed the MER for the sake of qualifying for a higher production premium. They also have to worry about restraining their geologists from drilling too many of the wrong kind of holes. Given the Soviet planning and incentive system, the geologists are pushed to drill as many holes as they can regardless of whether such drilling leads to the discovery of new petroleum. Because they work within yearly and five-year time-plan horizons, they have very short time-preferences. The geologists, along with the pumping crews, want to do whatever it takes to win premiums this year or in this five-year plan, not off in some distant and optimal but perhaps unrealizable future. Inevitably in the short run there is overdrilling and overpumping.

The Soviet incentive system is also the cause of their inadequate petroleum technology. It does not reward innovation or product adaptability. Consequently their drilling and recovery equipment, while useful for some purposes, is not as versatile as that in the West. This severely hampers not only their onshore and offshore drilling, but their efforts at enhanced recovery.

Ideological hang-ups also prevent them from properly reflecting the opportunity cost of using petroleum for domestic purposes. Since existing prices and profits of petroleum do not reflect the true value of a ton of increased petroleum production, the petroleum industry appears to be less profitable than it really is and this tends to make it a less likely recipient of new investment funds. As a result, allocation of investment is distorted. Utilization of shadow prices can alleviate this confusion somewhat, but not completely.

Finally, because ideology has lulled Soviet planners into believing they would be spared the effects of a recession, when the world economy began to falter in 1974 Soviet importers assumed that there was no need to take any defensive precautions. When they belatedly discovered their mistake, one of the few ways they could pay their import bills was to increase their export of petroleum. This led them to increase the export and the output of petroleum more than they would have preferred and more than was good for the long-run output of their wells.

POTENTIAL

Along with their problems, the Soviets also have potential. They have the largest sedimentary basins in the world, most of them only barely explored. Because their drilling operations are inadequate for the task, many of the fields they have explored and exploited still contain considerable supplementary potential. In some cases, what is needed is deeper or offshore drilling. In other cases, the Soviets could improve output considerably with secondary recovery methods or more economically rational procedures. Often all that is needed to take advantage of this potential is the purchase of foreign technology. The Soviets have recently come to recognize the need not only for this foreign technology but in several instances for foreign firms to assist them in joint ventures within the Soviet Union on- and offshore.

If they run out of petroleum, they still have considerable room to maneuver because they also have abundant alternative sources of energy. Their reserves of gas and coal are enormous. While there are problems of distance and technology to be conquered, the challenge is not insurmountable. Unlike we in the West, the Soviet are also moving rapidly ahead in expanding the use of nuclear energy. Finally, because the Soviets have considerable waste in their system they have the opportunity to improve the efficiency with which they consume their energy and raw materials. While there may not be much they can do to improve their already efficient public transportation system, there are other areas such as the use of energy in industry which are quite wasteful. But even with this potential, until recently, the Soviets have been unable to move resolutely. In part this was because in similar past crises new fields had always been discovered just as the older fields were being phased out. With such a large country, it seemed safe to assume that new fields would continue to be found in the future as well. Thus there was no need to worry too much about waste in the existing planning and incentive system. Soviet officials have always given lip service to the need to improve efficiency, but in the circumstances it was not too surprising that efforts at reforming the planning and incentive system, despite the evident waste, had never been too successful.

Any effective reduction in the long-accepted pattern of inefficiency and waste would require a fundamental change in the planning and incentive system and that would only be set off by a major jolt to the long-established perceptions of senior Soviet officials.

WHY THE CIA REPORT?

Granted that the difficulty in altering the Soviet system made it seem quite unlikely that there would be any immediate solution to the Soviet energy problem; yet with all the potential that is available to the Soviets, why did the CIA in its 1977 report concentrate only on the pessimistic side? Originally it allowed for only two options, the worst and the least worst. In the months that followed, we saw how the CIA gradually softened its position until in 1979 it conceded that the Soviet Union would probably not import any petroleum in 1985. But to protect itself, one would also have expected that the CIA would have also included a 'best case' scenario in its initial April 1977 report. This was also the conclusion of the staff report of the Senate's Select Committee on Intelligence of the United States Senate.[2] In other words, in its original report the CIA assumed that the Soviets would not be able to realize any of the benefits from 'the potential' they had. The CIA ignored the possibility of enhanced recovery of foreign technology, of joint ventures, of offshore drilling, of improved incentives and prices at home, and of the impossibility of paying so much hard currency for the projected imports. Why did the CIA issue this kind of report when it did?

EARLIER PREDICTIONS

It should be remembered that the CIA does not have a perfect record when it comes to predicting Soviet energy prospects. The CIA has been anticipating serious problems for the Soviet petroleum industry since the early 1970s. At best its predictions, if not wrong, have been consistently premature. That may be an unfair judgement to make. Who, including this author, has not erred in predicting some future Soviet action? None the less, for at least a decade, whether it was because the CIA failed to anticipate that the fourfold increase in prices in 1973 would make profitable much that had been unprofitable, or that Soviet workers would tolerate intolerable working conditions, or that output would be borrowed from the future in the expectation that other energy sources would be found, or that foreign technology could be imported to offset the poor quality of domestic equipment, the CIA's predictions about Soviet petroleum prospects have almost always erred on the negative side.

As far back as 1970, J. Richard Lee, one of the CIA's most respected analysts, expressed considerable skepticism about the prospects for the Soviet petroleum industry. Lee's essay was not released as an official statement by the CIA, yet its appearance in a compendium of papers published by the Joint Economic Committee called *Economic Performance and the Military Burden in the Soviet Union* required clearance from the CIA.[3] After quoting from P. D. Shashin, the then Minister of the Petroleum Industry, who stated that Soviet petroleum exports would not increase because of the rising domestic demand for oil, Lee went on to speculate that 'Soviet exports (of petroleum) to the free world may continue to decline along with earnings of hard currency from such exports.'

Three years later, writing in another compendium published by the Joint Economic Committee just before the 1973 oil embargo, Lee repeated his earlier doubts. Pointing out that the Soviet Union had urged its East European dependants 'to seek additional supplies of oil from the Middle East after 1975,' Lee concluded that 'the Soviet Union may not be able to provide enough oil to meet all the increased needs of the East Europeans, those of its expanding economy and still maintain exports to non-communist countries at the present levels.'[4]

Similar comments are encompassed in a formal agency statement released in mid-1974. 'The USSR has little, if any, uncommitted oil from domestic sources with which to expand sales to the West and take advantage of the present prices.'[5] That same year in testimony before the Joint Economic Committee, William Colby, the director of the CIA, asserted that 'exports of crude oil to the West are expected to peak in 1976 and then decline as a result of constraint on domestic supply, higher domestic consumption and commitment to Eastern Europe as well as limits on the volume of barter imports from the Middle East.'[6] Thereafter such statements became a ritual at the annual hearings conducted by the Joint Economic Committee for the CIA.[7] Only in 1976 was there any acknowledgement that the Soviets might be trying to solve their problems.[8]

How accurate were these predictions? While CIA staff members started to predict in 1970 that petroleum exports would not increase, actual net exports rose in every year but 1972 and gross exports rose in every year but 1974. When exports did not peak in 1976, the CIA again revised its estimates and pushed the expected peak to 1978.[9] The same predictions have been made and of necessity extended about production. In the April 1977 report, it was predicted that 'production will soon peak possibly as early' as 1978, and when it did not the decline was pushed off until 1979 and then 1980.[10]

When questioned about their over-pessimism, the CIA specialists insist that they would have been correct if the Soviets had not resorted to overpumping their deposits and borrowing from future production reserves. The CIA similarly concedes that initially it did not allow for the import of 'desperately needed Western technology, particularly submersible pumps and drill bits'.[11] Nor, as we saw, did the initial CIA report allow for the astronomical balance of trade deficit or the possibility of increased conservation. In other words, the CIA has generally assumed that the Soviet Union would be unresponsive to its problems.

Nor did the CIA anticipate the reaction to its report by Soviet planning officials. The CIA report had the effect of becoming a self-defeating prophecy. For years Soviet officials had been discussing the need to cope with their petroleum production problems but did little. It cannot be demonstrated conclusively but there are indications that the 1977 CIA report proved to be a major factor in prodding the Soviet authorities into action.

Apparently the CIA report was the answer to the Minister of the Petroleum Industry's prayers. Since the early 1970s, he had been pleading in vain for the hard currency to buy some of the more innovative technology imports. Instead he was largely restricted to money for pipeline construction equipment, drill bits, and submersible pumps. Part of his difficulty was that he lacked the documentation to prove how urgent his needs were. As we saw, because of ideological taboos, Soviet economic indicators were sending out incorrect signals. According to the official data, all Soviet energy industries appeared to be experiencing a falling rate of profit. This in turn strengthened those who insisted that Soviet investment and import resources should be put to better use elsewhere. In contrast, the CIA report seemed to verify everything the Minister of the Petroleum Industry had been saying. Moreover it was said by the CIA and the Soviets have traditionally paid more attention to what foreigners say than to their own specialists.

Since we do not have access to the discussions within Gosplan, Gosbank, and the Ministry of Finance, admittedly the impact of the CIA report can not be definitively proven. Nevertheless, remember, for example, that the Dresser drill bit plant had been under discussion since 1973. The Minister of the Petroleum Industry was all set to sign the contract but with evident embarrassment he confided to his American suppliers that he could not convince Gosplan, Gosbank, or the Ministry of Finance to allocate the necessary hard currency. Other Western and Japanese businessmen recount similar frustrations. However, once the April 1977 CIA report appeared, in rapid order the Soviets not only signed the contract for the drill bit plant, but also one for a factory to produce secondary recovery chemicals and a $228 million gas lift process to use instead of water injection in the west Siberian fields. This explains why overall purchases of petroleum technology in 1978 soared in comparison with previous years. According to a tabulation by *Business International*, sophisticated petroleum technology imports in 1978 rose by as much as two or three times the levels of previous years.[12] The release of this hard currency takes on all the more significance when it is remembered that in 1977, in other sectors of the Soviet economy, an effort was being made to curb hard currency imports. The release of money for petroleum technology was a major exception. From all appearances, therefore, it seems to have been the shock of the CIA prediction that finally broke through the ideological and pricing muddle that had stymied those who sought to bring in new technology for the Soviet petroleum industry. The change in attitude in late 1977 has not escaped the attention of the CIA. As their 1979 report put it, 'Soviet leaders have become aware of growing energy problems since at least the early 1970s, but did little about facing up to them until late 1977.'[13] What the CIA neglected to mention was that in large part that awakening was probably the result of its own April 1977 report. (If only the KGB would issue a report about the American energy problems!)

WHY WAS THE REPORT ISSUED?

It is impossible to know what exactly caused the CIA to issue the report when it did. There may have been some desire to support the urgency of President Carter's initial energy message, but a look at earlier statements by the CIA on the subject indicates the CIA said almost nothing in 1977 that it had not been saying repeatedly in one way or another since 1970. Thus the report was nothing more than an extension of earlier studies. In the circumstances it would have been surprising if the CIA had concluded that the Soviet Union would *not* have to import petroleum. As it was, the April 1977 report merely reflected an analytical consistency, not some research breakthrough. Yet while the difficulties the CIA has been focusing on all these years are real and compelling, they unduly reflect the engineering problems and ignore the economic and dynamic aspects of the analysis. Allowance should have been made for the fact that when the situation is urgent enough the Soviets will respond if need be by doing the unexpected, as they frequently have done in the past.

THE SWEDISH REPORT

While the Soviets have undoubtedly begun to rationalize some of their operating methods, it is unlikely that they will go so far as to make a thorough revamping of their overall planning and incentive system. Yet that is the assumption the so-called Swedish report seems to make.[14] For that reason the Swedish report is almost the mirror image of the April 1977 CIA report. Following Murphy's law, the CIA initially assumed that everything that could would go wrong. In contrast the Swedes seem to counter with what we here can call Swenson's law – that everything that could go well, would. As the Swedes see it, instead of the Soviet bloc having to import 3·5 to 4·5 MBD, the Soviet Union in 1985 will be able to export up to 185 million tons or 3·7 MBD to Western Europe. Considering that overall Soviet exports in 1979 totalled less than 160 million tons, of which only about 50 million went to the hard-currency countries, that will obviously be quite an accomplishment. Conceivably the Soviets could export that much to the hard-currency world, but only by ceasing exports to its allies and all other countries.

The reasoning underlying the Swedish report is that the Soviet planning and incentive system has led to enormous waste and under-utilization of existing wells.[15] With the proper change in incentives, older wells, particularly in the western part of the country, can easily increase their yields by 10 per cent to 35 per cent.[16] This higher yield can be obtained if the Soviets adopt a system of differentiated rent and marginal cost pricing.[17] In addition the Swedes also insist that a key finding by the CIA is incorrect. As we have seen, the fact that the water-cut in Samotlor has risen so fast means that output from Samotlor will probably fall shortly and sharply. The Swedes claim that the CIA is wrong about Samotlor. Citing a Soviet source, the Swedes assert that the water-cut in all of west Siberia in 1975 was only an average of 14·5 per cent.[18] The Swedes have apparently interpreted the CIA as saying that the water-cut at Samotlor in 1976 has increased to 57 per cent.[19] But if the water-cut level in all of west Siberia was 14·5 per cent, as the Swedes say, the comparable figure at Samotlor could not be 47 per cent since Samotlor production constitutes one-half of the total west Siberian output. This, according to the Swedes, constitutes a crucial flaw in the CIA argument. Instead, say the Swedes, the larger figure refers to Romashkino in the Volga-Ural region. If the Swedes are right, Samotlor's productive life is still vigorous and not likely to fall as quickly as the CIA predicts. Even if the field at Samotlor has 'peaked', they argue that this is not the same as saying the field is producing at its maximum capacity. As the Swedes see it, maximum producing capacity will not be realized for some years in the future.[20]

While some of the Swedish arguments are well taken, they do seem to err, at least partially, in their criticism of the CIA's prediction about the rising water levels at Samotlor. The CIA did not claim that the water level at Samotlor was 47 per cent, only that it was rising much faster than had been anticipated and faster than had been the experience in the Volga-Ural region. Moreover, the Swedish conclusions are so sweeping that they are suspect. Nor are the Swedes content with their prediction that hard-currency exports will total 3·7 MBD by 1985. They also predict that Soviet output will double by 1990![21] If true, that would be an incredible accomplishment. After all, we are not talking about insignificant quantities of petroleum. Much of the Swedish report is careful and cautious, but the jump to this basically unsubstantiated assumption raises all kinds of questions about the purpose of the report, particularly since it is difficult to find information about the authors and their sponsoring corporation, PetroStudies. Some American specialists have suggested that this might be a form of misinformation put out with at least the indirect blessing of the USSR in order to discredit the CIA. After all, the Soviets are having some real problems. Equally important, the likelihood of the type of economic reforms the Swedes insist the Soviets must have to increase efficiency will be slow in coming at best.

Whatever the merit of the Swedish analysis, it is the CIA report that has had the most impact on the United States and on policy-making. Not surprisingly, because of the CIA emphasis on the Soviet problems in producing petroleum, most Americans assume that the Soviet Union is an importer, not an exporter, of petroleum. It may have been unintentional, but the CIA has managed to confuse most Americans about Soviet energy. Remember that the 1977 CIA report, which attracted so much attention, came out only one year after the Soviets had managed to become the world's largest producer of petroleum. Yet instead of focusing on Soviet successes, the 1977 CIA report deflected the debate to the consideration of their problems. Not surprisingly, therefore, this influenced public and even governmental attitudes toward dealing with the USSR. In turn these attitudes affected debate over the sale of technology, especially petroleum technology, to the Soviet Union. One clear implication of the CIA report is that the Soviets desperately need our help and that there is no place for them to turn. This explains why critics such as Gershman base much of their argument for the strict control, if not embargo, on exports of petroleum technology on what they perceive to be the almost desperate Soviet need for technology that only we produce.[22] As a result some businessmen and other skeptics wonder whether or not a subtle purpose of the CIA report was to discourage technology exports to the Soviet Union.

LESSONS TO BE LEARNED

The CIA analysis of the Soviet petroleum industry is a good case-study of how a particularly competent and even impressive static analysis can be overtaken by the dynamics of the situation. Especially intriguing is the strong possibility that in part the dynamic change was made all the more effective by the reaction of Soviet policy-makers to the CIA's static analysis. Because the Soviet Union's own internal economic indicators in the petroleum industry had become distorted by ideological and doctrinal constraints, it took the CIA report to impress on Soviet leaders the seriousness of the situation which their own data tended to obscure.

But while the CIA report undoubtedly helped to clarify the situation and the real choices at stake in the Soviet Union, it had somewhat of an opposite effect in the United States. If anything, it tended to obfuscate the state of Soviet petroleum needs, and as a result it undoubtedly caused a certain amount of havoc in the way Americans view the scene. That raises the question of whether or not the CIA should have made its initial April 1977 findings public. After all, the President had a case to make for American energy conservation without bringing in the CIA analysis of the Soviet Union. In addition, making the report public in the way the CIA did focused enormous attention on the matter and as we saw was probably a factor in accelerating the Soviet effort to come to grips with its problems. The clearest indication that the Soviets have finally begun to respond is the article by V. Mishchevich, the First Deputy Minister of the Petroleum Industry, which was published in mid-1979.[23] In it he indicates that the Soviet government is taking exactly the kind of corrective steps that are needed to stop so much of the waste. For example, Mishchevich concentrates on the need to change the incentives of the drillers and geologists. He has proposed that drillers' premiums be awarded on the basis of productive wells actually opened up rather than on the number of meters drilled. Others have called for the reorganization of the energy industry to eliminate duplication and to ensure activity where none existed previously. Thus for many years there were three organizations responsible for drilling, but until 1979 there was no one organization charged with drilling offshore.

On a larger scale, in July 1979, a nationwide reform was announced which was intended to eliminate the excessive consumption of raw material inputs including energy.[24] Previously many Soviet industrial managers operated under a system whereby their quantitative targets were based on the 'gross value' of output produced. Under such a system, there were no incentives for Soviet managers to reduce or economize on the use of inputs. On the contrary, they were more likely to earn their premiums if raw materials and energy were squandered. This helps explain why Soviet machinery is so bulky and why the Soviets have had such problems with miniaturization. The understated prices of Soviet raw materials did nothing to counter such waste. The 1979 reform calls for the abolition of the 'gross system' and its replacement with a 'net normative' index. Thus Soviet managers will be judged henceforth according to the value added and therefore they presumably will find it no longer to their advantage to build up the cost of their production. There are also reports that a wholesale price increase will soon be forthcoming, the first in about thirteen years. This should further the incentive to conserve petroleum inside the USSR and thus increase the amount that can be set aside for export to Eastern and Western Europe.[25]

Conceivably, if the CIA report had not been written, the Soviets would not have been prodded into action as soon as they were and their economic difficulties would have become much more serious than is now likely to be the case. There is an element of truth to such an assertion, but it does not follow that the CIA should have classified its report. First, the CIA was not saying anything in its April 1977 report that it had not published previously. It was just that the April 1977 report, incorporated as it was into the President's message, made a bigger splash. Secondly, on major analytical matters of this sort which rely more on research and analysis than on covert operations, the CIA's work benefits from having an open review by outsiders. This should improve the quality of the ultimate analysis. Finally, even though an improvement in the Soviet energy situation will undoubtedly increase Soviet economic strength, this seems to be one area of activity where benefits to the Soviet Union also mean benefits to the United States. Unlike agriculture (a zero sum game) where a good crop in the Soviet Union is likely to mean fewer profits for American farmers, lower Soviet petroleum output is likely to mean more competition for American buyers of petroleum in the world market (a non-zero sum game), either because the Soviets will be unable to maintain their exports or because they may have to import petroleum from the world market themselves. Thus it would seem to be in our long-run interest to facilitate Soviet petroleum production, despite the fact that benefits will also accrue to the USSR.

There should be a lesson in this for all of us. For the Soviets, the moral is that if there is to be a basic improvement in the way they utilize their natural resources, and petroleum in particular, they will have to overcome some of their ideological and doctrinal hang-ups and introduce a new set of economic stimuli. They cannot always depend on the CIA to come to their rescue.

There is a different kind of moral for the United States, and it has to do with the scope of our intelligence work. Henceforth we should encourage the CIA to continue to expose analysis of this sort to public scrutiny. Equally important, when dealing with something as problematic as prospective energy production in the Soviet Union, the CIA should not let itself appear as if it has a stake in some particular outcome or that it has become a victim of its own consistency. Furthermore the CIA should be urged not to disregard evidence which leads to conclusions contrary to its own. It is important that the CIA's range of possible outcomes include not only a worst-case but also a best-case scenario. To focus only on one extreme may lead not only to a misguided set of policies for handling the problem but to a loss of credibility for the CIA.

NOTES: CHAPTER 8

1   *Pravda*, 30 December 1978, p. 2; *Sotsialisticheskaia industriia*, 20 October 1978, p. 2; *EKO*, March 1979, p. 14.
2   Senate Select Committee on Intelligence, the United States Senate, *The Soviet Oil Situation: An Evaluation of CIA Analyses of Soviet Oil Production* (Washington, DC: US Government Printing Office, 1978), p. 3.
3   The Joint Economic Committee, *Economic Performance and the Military Burden in the Soviet Union*, a Compendium of Papers (Washington, DC: US Government Printing Office, 1970), p. 35.
4   J. Richard Lee, 'The Soviet petroleum industry: promise and problems', *Soviet Economic Prospects for the 1970s*, the Joint Economic Committee (Washington, DC: US Government Printing Office, 27 June 1973), p. 287.
5   CIA, *The Soviet Economy in 1973* (Washington, DC, July 1974), p. 9.
6   Joint Economic Committee, *Allocation of Resources in the Soviet Union and China*, Hearings before the Sub-Committee on Priorities and Economy in Government, the Joint Economic Committee, Congress of the United States (Washington, DC: US Government Printing Office, 12 April 1974), p. 24 (hereafter this series is referred to as *Allocation of Resources*, JEC, and the appropriate date and page).
7   ibid., 1975, p. 9, 1977, pp. 5, 6, 48, 52, 53, 1978, pp. 4, 7, 80, 81.
8   ibid., 1976, p. 6.

9   CIA, *The Soviet Economy in 1976–77 and the Outlook in 1978* (Washington, DC, August 1978), p. 19.
10   CIA Soviet, April 1977, p. 9; *Allocation of Resources*, JEC, 1978, p. 4.
11   Staff Report of the Select Senate Committee on Intelligence of the US Senate, *The Soviet Oil Situation, An Evaluation of CIA Analyses of Soviet Production* (Washington, DC: US Government Printing Office, 1978), p. 11.
12   *Review of Sino-Soviet Oil*, August 1979, p. 65.
13   CIA, August 1979, p. 39.
14   PetroStudies, *Soviet Preparation for Major Oil Exports* (Malmo, Sweden, 1978).
15   ibid., p. 4.
16   ibid., pp. 6, 13.
17   ibid., p. 21.
18   ibid., p. 50; O. B. Kuznetsova and A. I. Zhechkov, 'Ekonomicheskaia effektivnost' razmeshcheniia dobychi nefti', *Ekonomika neftianoi promyshlennosti*, March 1978, pp. 8–9.
19   ibid., p. 50.
20   ibid., p. 15.
21   ibid., p. 38.
22   Gershman, p. 43.
23   *Sotsialisticheskaia industriia*, 27 July 1979, p. 2.
24   *Pravda*, 29 July 1979, p. 1.
25   *Ekonomicheskaia gazeta*, no. 35, August 1979, p. 5.

Appendix 1

Charter of the All-Union Foreign Trade Association Sojuznefteexport (V/O Sojuznefteexport)

I General

1. The All-Union Foreign Trade Association Sojuznefteexport, hereinafter referred to as V/O Sojuznefteexport, shall be a single independent economic complex operating on a profit-and-loss basis, fulfilling the duties placed on it and enjoying the rights pertaining to these activities. V/O Sojuznefteexport shall be a legal person.

2. V/O Sojuznefteexport shall operate on the basis of the Regulations on the All-Union Foreign Trade Association within the System of the USSR Ministry of Foreign Trade approved by Decision No. 416 of the Council of Ministers of the USSR of 31 May, 1978, and also on the basis of this Charter.

3. V/O Sojuznefteexport shall be liable for its obligations in that part of its assets which, under USSR legislation, is subject to attachment.

The state, its agencies and organisations shall not be liable for the obligations of V/O Sojuznefteexport, and V/O Sojuznefteexport shall not be liable for the obligations of the state, its agencies and organisations.

4. V/O Sojuznefteexport shall have a round seal inscribed with its name and emblem.

5. V/O Sojuznefteexport shall have its seat in Moscow at the postal address: 32/34, Smolenskaia Sq., Moscow, 121200; cable address: Moscow Nafta.

II. Firms and Other Organisations Constituting V/O Sojuznefteexport

6. V/o) Sojuznefteexport shall include the following specialised firms:

6.1. The Euronafta firm performing export and import operations in crude oil and petroleum products with organisations and firms in the countries of Western Europe.

6.2. The Internafta firm performing export and import operations in crude oil and petroleum products with organisations and firms in the countries of Eastern Europe and in other countries.

6.3. The Vostoknafta firm performing export and import operations in crude oil and petroleum products with organisations and firms in the countries of Africa, the Near and the Middle East.

6.4. The Dalnafta firm performing export and import operations in crude oil and petroleum products with organisations and firms in the countries of the Far

East and America, as well as export and import operations in petroleum coke, paraffin, benzene, toluene, ozokerite, kerosene, montan wax, oils and lubricants and other goods.

7. V/O Sojuznefteexport shall include the following transport and forwarding offices performing transport and forwarding operations with outward and inward cargoes, as well as other operations pertaining to the export and import of goods belonging to the product range of V/O Sojuznefteexport:

Office in the port of Novorossiisk;
Office in the port of Odessa;
Office in the port of Tuapse;
Office in the port of Batumi;
Office in the port of Ventspils;
Office in the port of Klaipeda;
Office in the port of Nakhodka.

8. Firms constituting V/O Sojuznefteexport shall not be legal persons and shall be guided in their activities by the regulations approved by the General Director of V/O Sojuznefteexport.

The firms shall be authorised to conclude, on the instructions and on behalf of V/O Sojuznefteexport, foreign trade transactions, issue, on behalf of V/O Sojuznefteexport, orders, in the established manner, to the suppliers of goods for export and also confirm acceptance of fulfilment of the assignments to import goods.

Offices constituting V/O Sojuznefteexport shall not be legal persons and shall be guided in their activities by the regulations approved by the General Director of V/O Sojuznefteexport.

III. Subject Matter and Objectives of Activities of V/O Sojuznefteexport

9. Objectives of the activities of V/O Sojuznefteexport shall be the fulfilment of plans, approved in the established manner, for the export and import of crude oil and petroleum products, the increased export of these goods, improvement of their quality and competitiveness, and increased effectiveness of the operations performed.

10. In accordance with the objectives of its activities V/O Sojuznefteexport shall:

10.1. perform operations to export from and import to the USSR crude oil and petroleum products;

10.2. render and accept services of a foreign trade nature related to the export from and import to the USSR of goods specified in Para. 10.1. of this Charter;

10.3. improve planning and the profit-and-loss balance, elaborate draft five-year and annual plans and present them for consideration in the established manner;

10.4. elaborate measures to develop new forms of foreign economic relations, in particular, agreements on a compensatory basis, production cooperation and barter operations;

10.5 elaborate and implement measures aimed at enhancing requirements of quality of imported goods;

10.6 study and make use of the situation on the relevant product markets;

10.7. elaborate and implement measures together with the relevant organisations with the aim of increasing the export of goods and improving its pattern, expanding the range, and enhancing the quality and competitiveness of exported goods;

10.8. elaborate and put into practice advertising measures with a view to expanding the export of goods within the range of V/O Sojuznefteexport;

10.9. elaborate and put into practice measures aimed at improving the transportation of relevant goods by rail, water, pipeline, motor and air transport, and at the increased and more effective use of national transport;

10.10. participate in the elaboration of measures to improve the operation of oil depots, ports, transport organisations effecting delivery, storage, loading and transportation of goods within the range fixed to V/O Sojuznefteexport;

10.11. undertake comparative analysis of domestic and similar foreign oil products, participate in the elaboration of standards and technical specifications for exported and imported petroleum products;

10.12. inform the relevant industrial ministries and departments of the world market requirements on the quality of goods, on goods in high demand, and on the measures taken by foreign firms to increase the competitiveness of goods;

10.13. ensure legal protection of its interests;

10.14. manage the activities of firms and other organisations constituting it;

10.15. elaborate and implement measures aimed at the economical expenditure of material resources and money.

IV. Powers of V/O Sojuznefteexport

11. To achieve the objectives specified in Para. 9 of this Charter, V/O Sojuznefteexport shall, in the manner established by USSR legislation, be authorised:

11.1. to perform, both in the USSR and abroad, all kinds of transactions and other legal deeds, including those of purchase and sale, barter, contract, loan, transportation, insurance, agency and commission, storage, joint activities and others, with institutions, enterprises, organisations, societies, partnerships and individuals, and also participate in tenders and competitions, and give guarantees;

11.2. to build, acquire, alienate, lease and rent, both in the USSR and abroad, enterprises auxiliary to its activities as well as all kinds of movable and immovable property;

11.3. to establish, both in the USSR and abroad, its affiliates, offices, branches, representations and agencies as well as establish or participate in all kinds of organisations whose activities comport with its tasks;

11.4. to sue and be sued in courts of law and arbitration and conclude amicable settlements;

11.5. to participate in international fairs and exhibitions and also participate in or organise specialised exhibitions, symposia, and publish advertising literature;

11.6. to ensure the participation of responsible representatives of relevant ministries and departments of the USSR in negotiations on the conclusion of major contracts for the export and import of goods specified in Para. 10.1. of this Charter.

V. *Management of V/O Sojuznefteexport*

Board of V/O Sojuznefteexport

12. The Board shall concentrate its activities on the fulfilment of plans, increased effectiveness of foreign trade operations, fulfilment by the suppliers of assignments on the export of goods set for them, fuller use of favourable situations on relevant product markets, realisation of organisational and technical measures to increase the export of above goods, expansion of their range, improvement of their quality and competitiveness, higher requirements of quality and technological standards of imported goods.

13. The quantitative and personal composition of the Board shall be approved by the Minister of Foreign Trade in agreement with the heads of the relevant industrial ministries and departments of the USSR.

Proposals on the quantitative and personal composition of the Board, and also, in appropriate cases, on changes in its composition shall be prepared by the General Director of V/O Sojuznefteexport.

14. The General Director of V/O Sojuznefteexport shall be the Chairman of the Board.

In case of the General Director's absence, fulfilment of the duties of the Chairman of the Board shall be placed on the Acting General Director of V/O Sojuznefteexport.

The Board shall elect the deputy (deputies) of the Chairman of the Board. Election or re-election of the deputy (deputies) of the Chairman of the Board shall be held at the first meeting of the Board during the current calendar year.

15. For the preparation of draft decisions and the organisation of fulfilment of the decisions made as well as for the proper organisation of control over the fulfilment of decisions made, the Board may establish relevant provisional working bodies accountable to the Chairman of the Board.

16. At its first meeting during the current calendar year the Board shall approve the annual plan of its work and shall also elaborate a plan of measures to organise the fulfilment of its decisions.

17. Meetings of the Board shall be held in accordance with the approved annual plan of the Board's work but not less than once a quarter.

Extraordinary meetings may be held at the proposal of not less than one-third of the members of the Board. An extraordinary meeting shall be held within the period proposed by the concerned members of the Board. This period shall not be shorter than ten working days starting from the date of introduction of the corresponding proposal.

18. Each member of the Board shall have one vote.

Decisions of the Board shall be taken by a majority of three-fourths of the votes of the attending members of the Board. Decisions of the Board shall be considered valid provided that the meeting of the Board is attended by not less than 50 per cent of members of the Board from organisations within the system of the Ministry of Foreign Trade, and by not less than 50 per cent of members of the Board from organisations of relevant industrial ministries and departments of the USSR.

19. Decisions of the Board shall be binding and shall be implemented through the orders of the General Director of V/O Sojuznefteexport.

20. The Board shall have a secretary who shall undertake the office work of

the Board and shall control the fulfilment of the decisions made by the Board.

21. Agreed minutes of the Board's meetings shall be signed by the Chairman and the Secretary of the Board; copies of agreed minutes shall be sent to the members of the Board within three days of the date of the Board's meeting.

General Director of V/O Sojuznefteexport

22. V/O Sojuznefteexport shall be headed by the General Director acting on the basis of undivided authority.

23. The General Director shall organise the work of V/O Sojuznefteexport and bear personal responsibility for all its activities.

The General Director shall manage operations of V/O Sojuznefteexport, including the solution of problems pertaining to the conclusion of contracts, act without a power of attorney on behalf of V/O Sojuznefteexport, represent it in all institutions, enterprises and organisations both in the USSR and abroad, administer the assets of V/O Sojuznefteexport in accordance with the USSR legislation and this Charter, perform all kinds of transactions and other legal deeds, issue powers of attorney, and open settlement and other accounts of V/O Sojuznefteexport with banks.

24. The General Director shall have deputies.

The competence of the General Director's deputies and other executives of V/O Sojuznefteexport shall be established by the General Director.

The competence of the Chief Accountant and the Head of the Legal Department of V/O Sojuznefteexport shall be determined by USSR legislation in force.

25. Regardless of the personal responsibility of the General Director for the activities of V/O Sojuznefteexport on the whole, the General Director's deputies and other executives of V/O Sojuznefteexport, as well as directors of the firms constituting it, shall bear responsibility for the state of operations in the respective fields.

VI. Signature of Foreign Trade Transactions on Behalf of V/O Sojuznefteexport

26. Foreign trade transactions concluded by V/O Sojuznefteexport shall be signed by two persons.

Authority to sign such transactions shall belong to the General Director of V/O Sojuznefteexport, his deputies, directors of firms constituting V/O Sojuznefteexport, and also to the persons authorised by a power of attorney signed by the General Director of V/O Sojuznefteexport.

27. Bills of exchange and other monetary obligations issued by V/O Sojuznefteexport must be signed by two persons – the General Director of V/O Sojuznefteexport or one of his deputies and the Chief Accountant of V/O Sojuznefteexport or a person authorised by a power of attorney signed by the General Director and the Chief Accountant of V/O Sojuznefteexport.

28. The General Director of V/O Sojuznefteexport shall determine the values and types of transactions whose signature, in accordance with the procedure established in Paragraphs 26 and 27 of this Charter, shall belong to the competence of his deputies and directors of the firms constituting V/O Sojuznefteexport.

VII. *Assets of V/O Sojuznefteexport*

29. The assets of V/O Sojuznefteexport shall consist of fixed and current assets constituting its authorised fund, as well as of funds for material stimulation, social and cultural functions and housing construction, development of foreign trade activities, and temporary financial assistance and of other assets.

The assets of V/O Sojuznefteexport shall be reflected in its balance drawn up in the manner established by USSR legislation.

The limits and rules of establishing and using the above funds shall be determined by USSR legislation.

30. V/O Sojuznefteexport shall have an authorised fund of 5 (five) million rubles.

VIII. *Accounting and Distribution of Profits of V/O Sojuznefteexport*

31. The trading year of V/O Sojuznefteexport shall be from 1 January to 31 December of the calendar year.

32. The accounts of V/O Sojuznefteexport shall be drawn up and approved in the manner established by USSR legislation.

The draft annual report of V/O Sojuznefteexport shall be considered at the meeting of the Board of V/O Sojuznefteexport.

33. The rules governing the distribution of profits of V/O Sojuznefteexport shall be determined by USSR legislation.

34. Auditing of financial and economic activities of V/O Sojuznefteexport shall be held in accordance with USSR legislation not less than once a year.

IX. *Reorganisation and Liquidation of V/O Sojuznefteexport*

35. Reorganisation and liquidation of V/O Sojuznefteexport shall be implemented in accordance with USSR legislation.

Appendix 2

Charter of the All-Union Foreign Trade Association Sojuzgazexport (V/O Sojuzgazexport)

I. General

1. The All-Union Foreign Trade Association Sojuzgazexport, hereinafter referred to as V/O Sojuzgazexport, shall be a single independent economic complex operating on a profit-and-loss basis, fulfilling the duties placed on it, and enjoying the rights pertaining to these activities. V/O Sojuzgazexport shall be a legal person.

2. V/O Sojuzgazexport shall operate on the basis of the Regulations on the All-Union Foreign Trade Association within the system of the USSR Ministry of Foreign Trade approved by Decision No. 416 of the Council of Ministers of the USSR of 31 May 1978, and also on the basis of this Charter.

3. V/O Sojuzgazexport shall be liable for its obligations in that part of its assets which, under USSR legislation, is subject to attachment.

The state, its agencies and organisations shall not be liable for the obligations of V/O Sojuzgazexport, and V/O Sojuzgazexport shall not be liable for the obligations of the state, its agencies and organisations.

4. V/O Sojuzgazexport shall have a round seal inscribed with its name and emblem.

5. V/O Sojuzgazexport shall have its seat in Moscow at the postal address: 20, Lenin Prospekt, Moscow. Cable address: Moscow Sovgas.

II. Firms Constituting V/O Sojuzgazexport

6. V/O Sojuzgazexport shall include the following specialised export-import firms:

6.1. The Eximgas firm for trade in natural gas with the countries of Western Europe, America and Asia.

6.2. The Intergas firm for trade in natural gas with the countries of Eastern Europe.

6.3. The Spetsgas firm for trade in liquefied oil and special gases.

6.4. The Transitgas firm for the transit of natural gas.

7. Firms constituting V/O Sojuzgazexport shall not be legal persons and shall be guided in their activites by the regulations approved by the General Director of V/O Sojuzgazexport with reference to the Model Regulations on the Specialised Firm of an All-Union Foreign Trade Association.

The firms shall be authorised to conclude, on the instructions and on behalf of V/O Sojuzgazexport, foreign trade transactions, issue, on behalf of V/O Sojuzgazexport, orders, in the established manner, to the suppliers of goods for export and also confirm acceptance of fulfilment of the assignments to import goods.

III. Subject Matter and Objectives of Activities of V/O Sojuzgazexport

8. Objectives of the activities of V/O Sojuzgazexport shall be the fulfilment of plans, approved in the established manner, for the export from and import to the USSR of natural, liquefied oil, and other gases, the increased export of relevant goods, improvement of their quality and competitiveness, and increased effectiveness of the operations performed.

9. In accordance with the objectives of its activites V/O Sojuzgazexport shall:

9.1. perform operations to export from and/or import to the USSR goods within the relevant range;

9.2. render and accept services of a foreign trade nature related to the export from and/or import to the USSR of relevant goods;

9.3. improve planning and the profit-and-loss balance, elaborate draft five-year and annual plans and present them for consideration in the established manner;

9.4. elaborate measures to develop new forms of foreign economic relations, in particular, agreements on a compensatory basis, production cooperation, and barter operations;

9.5. study and make use of the situation in the relevant product markets;

9.6. elaborate and put into practice advertising measures with a view to expanding the export of goods within its range;

9.7. elaborate and put into practice measures aimed at improving the transportation of relevant cargoes by rail, water, air, motor, and pipeline transport, and at the increased and more effective use of national transport;

9.8. participate in the elaboration of measures to improve the storage and ensure the safe-keeping of relevant cargoes;

9.9. undertake comparative analysis of new domestic and analogous foreign technology within the relevant product range and data on gas recovery, production and transportation;

9.10. inform the relevant industrial ministries and departments of the world market requirements on the quality of goods, on goods in high demand, and on the measures taken by foreign firms to increase the competitiveness of goods;

9.11. ensure legal protection of its interests;

9.12. manage the activities of firms constituting it;

9.13. elaborate and implement measures aimed at the economical expenditure of material resources and money.

IV. Powers of V/O Sojuzgazexport

10. To achieve the objectives specified in Para. 8 of this Charter, V/O Sojuzgazexport shall, in the manner established by USSR legislation, be authorised:

10.1. to perform both in the USSR and abroad all kinds of transactions and other legal deeds, including those of purchase and sale, barter, contract, loan, transportation, insurance, agency and commission, storage, joint activities, and others, with institutions, enterprises, organisations, societies, partnerships and individuals, and also participate in tenders and competitions, and give guarantees;

10.2. to build, acquire, alienate, lease and rent, both in the USSR and abroad, enterprises auxiliary to its activities, as well as all kinds of movable and immovable property;

10.3. to establish its affiliates, offices, branches, representations, and agencies, as well as establish or participate in all kinds of organisations whose activities comport with its tasks;

10.4. to sue and be sued in courts of law and arbitration, and conclude amicable settlements;

10.5. to participate in international fairs and exhibitions, and also participate in or organise specialised exhibitions, symposia, and publish advertising literature;

10.6. to ensure the participation of responsible representatives of relevant ministries and departments of the USSR in negotiations on the conclusion of major contracts for export and/or import of goods.

V. Management of V/O Sojuzgazexport
Board of V/O Sojuzgazexport

11. The Board shall concentrate its activities on the fulfilment of plans, increased effectiveness of foreign trade operations, fulfilment by the suppliers of assignments on the export of goods allocated to them, fuller use of favourable situations in relevant product markets, realisation of organisational and technical measures to increase the export of the above goods, expansion of their range, improvement of their quality and competitiveness, higher requirements of quality and technological standards of imported goods.

12. To fulfil the tasks specified in Para. 11 of this Charter, the Board shall:

12.1. review draft five-year and annual plans and the implementation of these plans;

12.2. review issues pertaining to the state and use of situation in the markets;

12.3. take decisions on the elaboration of organisational and technical measures to increase the export of goods, expand the range, increase the quality and competitiveness of exported goods and the effectiveness of foreign trade operations;

12.4. resolve matters related to finding new goods for export and those over and above the plan, and introduce proposals in the established manner;

12.5. study questions of specialisation of enterprises in the production of goods for export and elaborate relevant proposals;

12.6. consider matters related to increasing the requirements of quality and technological standards of goods imported and exported by the Association, and elaborate, whenever necessary, relevant proposals;

12.7. consider matters on the development of new forms of foreign economic relations;

12.8. effect control over the fulfilment by the suppliers of set assignments and orders issued by the Association for the delivery of goods for export;

12.9. consider questions of further improving the activities of the firms constituting the Association;

12.10 consider matters related to claims about the quality of exported and imported goods, and elaborate the relevant measures.

13. The quantitative and personal composition of the Board shall be approved by the Minister of Foreign Trade in agreement with the heads of the relevant industrial ministries and departments of the USSR.

Proposals on the quantitative and personal composition of the Board, and also, in appropriate cases, on changes in its composition shall be prepared by the General Director of V/O Sojuzgazexport.

14. The General Director of V/O Sojuzgazexport shall be the Chairman of the Board.

In case of the General Director's absence, fulfilment of the duties of the Chairman of the Board shall be placed on the Acting General Director of V/O Sojuzgazexport.

The Board shall elect the deputy of the Chairman of the Board. Election or re-election of the deputy of the Chairman of the Board shall be held at the first meeting of the Board during the current calendar year.

15. For the preparation of draft decisions and the proper organisation of control over the fulfilment of decisions made, the Board may establish relevant provisional working bodies accountable to the Chairman of the Board.

16. At its first meeting during the current calendar year the Board shall approve the annual plan of its work and shall also elaborate a plan of measures to organise the fulfilment of its decisions.

17. Meetings of the Board shall be held in accordance with the approved annual plan of the Board's work but not less than once a quarter.

Extraordinary meetings may be held at the proposal of not less than one-third of the members of the Board. An extraordinary meeting shall be held within the period proposed by the concerned members of the Board. This period shall not be shorter than ten working days, starting from the date of introduction of the corresponding proposal.

18. Each member of the Board shall have one vote. Decisions of the Board shall be taken by a majority of three-fourths of the votes of the attending members of the Board. Decisions of the Boards shall be considered valid provided that the meeting of the Board is attended by not less than 50 per cent of members of the Board from organisations within the system of the Ministry of Foreign Trade, and by not less than 50 per cent of members of the Board from organisations of relevant industrial ministries and departments of the USSR.

19. Decisions of the Board shall be binding and shall be implemented through the orders of the General Director of V/O Sojuzgazexport.

20. The Board shall have a Secretary who shall undertake the office work of the Board and shall control the fulfilment of the decisions made by the Board.

21. Agreed minutes of the Board's meetings shall be signed by the Chairman and the Secretary of the Board; copies of agreed minutes shall be sent to the members of the Board within three days of the date of the Board's meeting.

General Director of V/O Sojuzgazexport

22. V/O Sojuzgazexport shall be headed by the General Director acting on the basis of undivided authority.

23. The General Director shall organise the work of V/O Sojuzgazexport and bear personal responsibility for its activities.

The General Director shall manage the operations of V/O Sojuzgazexport, including the solution of problems pertaining to the conclusion of contracts, act without the power of attorney on behalf of V/O Sojuzgazexport, represent it in all institutions, enterprises and organisations both in the USSR and abroad, administer the assets of V/O Sojuzgazexport in accordance with USSR legislation and this Charter, perform all kinds of transactions and other legal deeds, issue powers of attorney, and open settlement and other accounts of V/O Sojuzgazexport with banks.

24. The General Director shall have deputies. The competence of the General Director's deputies and other executives of V/O Sojuzgazexport shall be established by the General Director.

The competence of the Chief Accountant and the Head of the Legal Department of V/O Sojuzgazexport shall be determined by USSR legislation in force.

25. Regardless of the personal responsibility of the General Director for the activities of V/O Sojuzgazexport on the whole, the General Director's deputies and other executives of V/O Sojuzgazexport, as well as directors of the firms constituting it, shall bear responsibility for the state of operations in the respective fields.

VI. Signature of Foreign Trade Transactions on Behalf of V/O Sojuzgazexport

26. Foreign trade transactions concluded by V/O Sojuzgazexport shall be signed by two persons.

Authority to sign such transactions shall belong to the General Director, his deputies, directors of firms, and also to the persons authorised by a power of attorney signed by the General Director of V/O Sojuzgazexport.

27. Bills of exchange and other monetary obligations issued by V/O Sojuzgazexport must be signed by two persons – the General Director of V/O Sojuzgazexport or one of his deputies and the Chief Accountant or a person authorised by a power of attorney signed by the General Director and the Chief Accountant of V/O Sojuzgazexport.

The said bills of exchange and other monetary obligations may also be signed by two persons authorised by powers of attorney signed by the General Director and the Chief Accountant of V/O Sojuzgazexport.

28. The General Director of V/O Sojuzgazexport shall determine the values and types of transactions whose signature, in accordance with the procedure established in Paras. 26 and 27 of this Charter, shall belong to the competence of his deputies and directors of firms.

VII. Assets of V/O Sojuzgazexport

29. The assets of V/O Sojuzgazexport shall consist of fixed and current assets constituting its authorised fund, as well as of funds for material

stimulation, social and cultural functions and housing construction, development of foreign trade activities, and temporary financial assistance, and of other assets.

The assets of V/O Sojuzgazexport shall be reflected in its balance drawn up in the manner established by USSR legislation.

The limits and rules of establishing and using the above funds shall be determined by USSR legislation.

30. V/O Sojuzgazexport shall have an authorised fund of 15 (fifteen) million rubles.

VIII. Accounting and Distribution of Profits of V/O Sojuzgazexport

31. The trading year of V/O Sojuzgazexport shall be from 1 January to 31 December of the calendar year.

32. The accounts of V/O Sojuzgazexport shall be drawn up and approved in the manner established by USSR legislation.

The draft annual report of V/O Sojuzgazexport shall be considered at the meeting of the Board.

33. The rules governing the distribution of profits of V/O Sojuzgazexport shall be determined by USSR legislation.

34. Auditing of financial and economic activities of V/O Sojuzgazexport shall be held in accordance with USSR legislation not less than once a year.

IX. Reorganisation and Liquidation of V/O Sojuzgazexport

35. Reorganisation and liquidation of V/O Sojuzgazexport shall be implemented in accordance with USSR legislation.

Bibliography

M. A. Adelman, *The World Petroleum Market* (Baltimore: Johns Hopkins University Press, 1972).

American Petroleum Institute, *Petroleum in the Soviet Union* (Washington, DC: mimeo., undated), p. 60.

R. M. Andreasian and A. D. Kaziukov, *OPEC Mire Nefti* (Moscow: Nauka, 1978).

N. K. Baibakov, 'Soglasovannyi plan mnogostornnikh integratsionnykh meropriiati – novaia stipen' razvitiia sovmestnoi planovoi deiatel'nosti stran – chlenov CEV', *Planovoe khoziaistvo*, September 1975.

D. V. Belorusov, I. I. Panfilov and V. A. Sennikov, *Problemy razvitiia i razmeshcheniia proizvoditel'nykh sil zapadnoi Sibiri* (Moscow: Mysl', 1976).

'Berech' i umnozhat' prirodnye bogatstva', *Planovoe khoziaistvo*, June 1973, p. 5.

D. V. Belorusov, I. I. Pafilov and V. A. Sennikov, *Problem' razvitiia i razmeshchenie proizvoditel'nykh sil zapadnoi Sibiri* (Moscow: Mysl', 1976).

C. Fred Bergsten, 'The threat is real', *Foreign Policy*, no. 14, Spring 1974.

Joseph S. Berliner, *The Innovation Decision in Soviet Industry* (Cambridge: MIT Press, 1976).

A. Beschinskii and R. Vitebskii, 'Energetika i razemeshchenie promyshlennogo proizvodstva', *Voprosy ekonomiki*, May 1977.

O. Bogomolov, 'Khoziaistvennye reformy i ekonomicheskoe sotrudnichestvo sotsialisticheskikh stran', *Voprosy ekonomiki*, February 1966.

Robert W. Campbell, *The Economics of Soviet Oil and Gas* (Baltimore: Johns Hopkins University Press, 1968).

Robert W. Campbell, *Soviet Energy Balances*, R-2257-DOE (Santa Monica: RAND, December 1978).

Robert W. Campbell, *Soviet Energy R & D: Goals, Planning, and Organizations*, R-2253-DOE (Santa Monica: RAND, May 1978).

Robert W. Campbell, *Trends in the Soviet Oil and Gas Industry* (Baltimore: Johns Hopkins University Press, 1976).

Central Intelligence Agency, *Handbook of Economic Statistics 1978: A Research Aid*, ER 78-10365 (Washington, DC: October 1978).

Central Intelligence Agency, *Prospects for Soviet Oil Production*, ER 77-10270 (Washington, DC: April 1977).

Central Intelligence Agency, *Prospects for Soviet Oil Production: A Supplemental Analysis*, ER 77-10425 (Washington, DC: July 1977).

Central Intelligence Agency, *Research Aid, Soviet Long-Range Energy Forecasts*, A(ER) 75-71 (Washington, DC: September 1975).

Central Intelligence Agency, *Simulations of Soviet Growth Options to 1985*, ER 79-10131 (Washington, DC: March 1979).

Central Intelligence Agency, *Soviet Commercial Operations in the West*, ER 77-10486 (Washington, DC: September 1977).

Central Intelligence Agency, *The Soviet Economy in 1976–77 and Outlook for 1978: A Research Paper*, ER 78-10512 (Washington, DC: August 1978).

Central Intelligence Agency, *Soviet Economic Problems and Prospects*, ER 77-10436U (Washington, DC: July 1977).

Central Intelligence Agency, *The World Oil Market in the Years Ahead*, ER 79-10327U (Washington, DC: August 1979).

Leslie Dienes, 'Energy policy changes and prospects in East Central Europe', mimeo.

Leslie Dienes, 'The Soviet Union: an energy crunch ahead?' *Problems of Communism*, September–December 1977.

Leslie Dienes and Theodore Shabad, *The Soviet Energy System: Resource Use and Policies* (New York: Halsted Press, Wiley, 1979).

Raimund Dietz, 'Price changes in Soviet trade with other CMEA countries and the rest of the world since 1975', in *Soviet Economy in a Time of Change*, vol. 1. Joint Economic Committee, US Congress (Washington, DC: US Government Printing Office, 10 October 1977), p. 263.

Iain F. Elliot, *The Soviet Energy Balance* (New York: Praeger, 1974).

'Estimate of the world's recoverable crude oil resources', *World Petroleum Congress*, PD 6 (2).

Europa-sibir, *Ekonomika i organizatsiia promyshlennogo proizvodstva* (EKO), no. 4, 1976.

Export-Import Bank Act, 1945, Washington, DC, Export-Import Bank.

N. K. Feitel'man, 'Ob ekonomicheskom otsenke mineral'nykh resursov', *Voprosy ekonomiki*, November 1968.

N. G. Feitel'man, 'Sotsial'no-ekonomicheskie problemy ekologicheskogo ravnovesiia v zapadnoi sibiri', *Voprosy ekonomiki*, October 1978.

P. H. Frankel, *Mattei: Oil and Power Politics* (New York: Praeger, 1966).

K. E. Gabyshev, 'Ekonomicheskaia otsenka prirodnykh resursov i rentnye platezhi', *Vestnik Moskovskogo universiteta, seriia ekonomika*, no. 5, 1969.

Carl Gershman, 'Selling them the rope: business and the Soviets', *Commentary*, April 1979.

Marshall I. Goldman, 'The relocation and growth of the pre-revolutionary Russian ferrous metal industry', *Explorations in Entrepreneurial History*, vol. 9, no. 1, October 1954.

Marshall I. Goldman, *The Spoils of Progress* (Cambridge: MIT Press, 1972).

John R. Haberstroh, 'Eastern Europe: growing energy problems', *East European Economics, Post-Helsinki*, Joint Economic Committee, US Congress (Washington, DC: US Government Printing Office, 25 August 1977).

John P. Hardt, Ronda A. Bresnick and David Levine, 'Soviet oil and gas in the global perspective', *Project Interdependence: US and World Energy Outlook through 1990* (Washington, DC: Congressional Research Library of Congress, November 1977).

Hearings before the Sub-Committee to Investigate the Administration of the

Internal Security Act and Other Internal Security Laws of the Committee on the Judiciary, United States Senate, *Exports of Strategic Materials to the USSR and Other Soviet Bloc Countries*; Eighty-Seventh Congress, Second Session (Washington, DC: US Government Printing Office, 1963), pt III.

Edward A. Hewett, 'The Soviet and East European energy crisis: its dimensions and implications for East–West trade', Center for Energy Studies, University of Texas at Austin, August 1978, mimeo.

Edward A. Hewett, 'Soviet primary product exports to CMEA and the West', American Association of Geographers, Washington, DC, *Project on Soviet Natural Resources in the World Economy*, no. 9, May 1979.

Samuel P. Huntington, 'Trade Technology and Leverage: Economic Diplomacy', *Foreign Policy*, no. 32, Fall 1978.

Iu. Iakovetz, 'Dvizhenie tsen mineral'nogo syr'ia', *Voprosy ekonomiki*, June 1975.

Iu. Iakovetz, 'Ekonomicheskie rychagi i povyshenie effektivnosti mineral'no-syr'evogo kompleksa', *Planovoe khoziaistvo*, January 1978.

L. Iofina, 'Prognozirovanie potrebnosti v energeticheskikh resursakh v ASPR Gosplana SSSR', *Planovoe khoziaistvo*, November 1977.

Neil H. Jacoby, *Multi-National Oil* (New York: Macmillan, 1974).

Robert G. Jensen (ed.), *Soviet Energy Policy and the Hydrocarbons: Comments and Rejoinders*, Washington, DC, American Association of Geographers, no. 7, February 1979.

Joint Economic Committee, *Allocation of Resources in the Soviet Union and China*, Hearings before the Sub-Committee on Priorities and Economy in Government, Joint Economic Committee, Congress of the United States (Washington, DC: US Government Printing Office, 12 April 1974).

Joint Economic Committee, *Economic Performance and the Military Burden in the Soviet Union*, a Compendium of Papers (Washington, DC: US Government Printing Office, 1970).

Joint Economic Committee, *Soviet Economic Prospects for the Seventies* (Washington, DC: US Government Printing Office, 27 June 1973).

T. Khachaturov, 'Ekonomicheskie problemy ekologii', *Voprosy ekonomiki*, June 1978.

T. Khachaturov, 'Prirodyne resursy i planirovanie narodnogo khoziaistva', *Voprosy ekonomiki*, August 1973.

T. Khachaturov and M. Loiter, 'Ekonomicheskaia tsenka prirodnykh resursov pri proektirovanii i stroitel'stve', *Vestnik AN SSSR*, March 1973.

Arthur Jay Klinghoffer, *The Soviet Union and International Oil Politics* (New York: Columbia University Press, 1977).

Boris Komarov, *Unichtozhenie prirody* (Frankfurt: Posev, 1978).

Ben Korda, 'Economic policies and energy production in Czechoslovakia', *ACES Bulletin*, vol. XVII, nos 2–3, Winter 1975.

Ben Korda and Ivan Moravcik, 'The energy problem in Eastern Europe and the Soviet Union', *Canadian-Slavonic Papers*, March 1976.

Barry Kostinsky, *Description and Analysis of Soviet Foreign Trade Statistics*, Foreign Demographic Analysis Division, Bureau of Economic Analysis, Social and Economic Statistics Administration, US Department of Commerce, Foreign Economic Reports, no. 5, 1974.

A. P. Krylov, 'O nekotorye voprosakh nefteotdachi', *Neftianoe khoziaistvo*, no. 3, 1974.

A. Lalaianitz, 'Dva obshchestva – gve linin v razvitii toplivo-energeticheskogo kompleksa', *Planovoe khoziaistvo*, April 1975.

J. Richard Lee, 'The Soviet petroleum industry: promise and problems', *Soviet Economic Prospects for the Seventies*, Joint Economic Committee, Congress of the United States (Washington, DC: US Government Printing Office, 1973).

S. Levin, V. Vasil'eva and N. Kosinov, 'Tsenoobrazovanie v neftianoi promyshlennosti', *Planovoe khoziaistvo*, July 1976, p. 111.

'List of foreign source drilling equipment for semi-submersible exploratory drilling vessel capable of drilling in six hundred feet of water to a depth of 25,000 feet', Washington, DC, Armco, 22 February 1979.

Peter Lyashchenko, *History of the National Economy of Russia to the 1917 Revolution* (New York: Macmillan, 1949).

G. I. Martsinkevich, *Ispol'zovanie prirodnykh resursov i okhrana prirody* (Minsk: BGU, 1977).

Ia. Mazover, 'Perspektivy Kansko-Achinskogo ugol'nogo basseina', *Planovoe khoziaistvo*, June 1975.

Ia. Mazover, 'Puti razvitiia toplivnogo khoziaistva SSSR', *Planovoe khoziaistvo*, November 1977.

Robert J. McIntyre and James R. Thornton, 'Urban design and energy utilization: a comparative analysis of Soviet practice', *Journal of Comparative Economics*, no. 4, December 1978.

N. Mel'nikov and V. Shelest, 'Toplivno-energeticheskii kompleks SSSR', *Planovoe khoziaistvo*, February 1975.

Ministerstvo vneshnei torgovli SSSR, *Vneshniaia torgovlia SSSR za 1918–1940* (Moscow: Vneshtorgizdat, 1960).

G. Mirlin, 'Effektivnost' ispol'zovaniia mineral'nykh resursov', *Planovoe khoziaistvo*, no. 6, 1973.

John D. Moody, 'World crude resources may exceed 1,500 billion barrels', *World Oil*, September 1975.

John D. Moody and Robert E. Geiger, 'Soviet petroleum resources: how much oil and where?', *Technology Review*, March/April 1975.

Donald L. Mueller, 'Siberian natural gas for the US', paper delivered at the Eastern AAASS Meeting in Storrs, Connecticut, 1 May 1976.

V. I. Muravlenko and V. I. Kremneva (eds), *Sibirskaia Neft'* (Moscow: Nedra, 1977).

A. M. Nekrasov, 'Osnovnye napravleniia ratsional'nogo ispol'zovaniia i ekonomii energoresursov', *Planovoe khoziaistvo*, May 1978.

A. M. Nekrasov and M. C. Pervukhin, *Energetika SSSR v 1976–1980 godakh* (Moscow: Energiia, 1977).

V. Novikov, 'K novomy pod'emu otchestvennogo mashinostroeniia', *Kommunist*, no. 3, February 1979.

E. Nukhovich, 'Ekonomicheskoe sotrudnichestvo SSSR s osvobodivshimisia stranami i burzhuaznye kritiki', *Voprosy ekonomiki*, October 1966.

M. M. Odintsov and A. A. Bykharov, 'Mineral'nye resursy zony', *Vestnik AN SSSR*, September 1975.

'Osnovnye problemy kompleknogo razvitiia zapadni sibiri materialy, tola', *Voprosy filosofii*, September 1978.

'Otsenka prirodnykh resursov', *Voprosy geografii* (Moscow: Mysl', 1968).

Daniel Park, *Oil and Gas in COMECON Countries* (London: Kogan Page, 1979).

Petro-Studies Co., *Soviet Preparations for Major Boost of Oil Exports* (Malmo, Sweden: PetroStudies Co., 1978).

V.I. Pokrovskii, *Sbornik svidenii po istorii i statistike vneshnei torgovli Rossii* (St Petersburg: Department tamazhennykh sborov, 1902), vol. I.

L. A. Potemkin, *Okhrana nedr i okruzhaiushchei prirody* (Moscow: Nedra, 1977).

A. Probst, 'Puti razvitiia toplivnogo khoziaistva SSSR', *Voprosy ekonomiki*, June 1971.

G. Prokhorov, 'Mirovaia sistema sotsializma i osvobodivshiesia strany', *Voprosy ekonomiki*, November 1965.

Jack H. Ray, 'The Role of Soviet Natural Gas in East–West Co-operation', Vienna, mimeo. prepared for Vienna II Conference, 5–8 March 1979.

Jeremy Russell, *Energy as a Factor in Soviet Foreign Policy* (Westmead, England, and Lexington, Mass.: Saxon House/Lexington Books, 1976).

Herbert Sawyer, 'The Soviet energy sector: problems and prospects', in *The USSR in the 1980s – Economic Growth and the Role of Foreign Trade* (Brussels: NATO, 1978).

N. S. Sazykin, *Mineral'no-syr'evye resursy* (Moscow: Zanie seriia nauka o zemble, June 1975).

G. Segal, *Oil and Petrochemical Industry in Soviet Union* (London: mimeo., undated).

M. Sladkovskii, 'XXII s'ezd KPSS i problemy ekonomicheskogo sotrudnichestva sotsialisticheskikh stran', *Voprosy ekonomiki*, April 1966.

Marianna Slocum, 'Soviet energy: an international assessment', *Technology Review*, October/November 1974.

Soviet ekonomicheskoi vzaimopomoshchi, Statisticheskii ezhegodnik 1971 (Moscow: Statistika, 1971).

Soviet Power Reactors, 1974, Report of the United States Nuclear Power Delegation Visit to the USSR, 19 September–1 October 1974, USERDA, ERDA-2 (Washington, DC, 1974).

Staff Report of the Select Senate Committee on Intelligence of the US Senate, *The Soviet Oil Situation; An Evaluation of CIA Analyses of Soviet Production* (Washington, DC: US Government Printing Office, 1978).

Jonathan P. Stern, *Soviet Natural Gas Development to 1990* (Lexington, Mass.: Lexington Books, 1980).

Jonathan P. Stern, *Soviet Natural Gas in the World Economy*, Washington, DC, Association of American Geographers, no. 11, 9 June 1979.

Robert Stobaugh and Daniel Yergin (eds), *Energy Future* (New York: Random House, 1979).

'The structural settings for giant gas and oil fields', panel discussion, Rome I (4).

Anthony C. Sutton, *Western Technology and Soviet Economic Development, Vol. I, 1917 to 1930* (Stanford: Hoover Institution Press, 1968).

Anthony C. Sutton, *Western Technology and Soviet Economic Development, Vol II, 1930 to 1945* (Stanford: Hoover Institution Press, 1971).

Antony C. Sutton, *Western Technology and Soviet Economic Development, Vol. III, 1945 to 1965* (Stanford: Hoover Institution Press, 1973).

Judith A. Thornton, 'Soviet methodology for the valuation of natural resources', *Journal of Comparative Economics*, no. 4, December 1978.

N. Tikhonov, 'Sovershenstovanie upravleniia-vazhnyi reserv povysheniia effektivnosti obshchestvennogo proizvodstva', *Kommunist*, no. 7, May 1979.

'Tiumen': Kompleke i ego grani', *Ekonomika i organizatsiia promyshlennogo proizvodstva*, no. 3, March 1979.

Robert W. Tolf, *The Russian Rockefellers* (Stanford: Hoover Institution Press, 1976).

Trudy statisticheskogo otdelenie, departament tamozhenny sbor', *Obzor' Vneshnei torgovli Rossii za 1901* (St Petersburg: Department tamozhenny sbor', 1903).

Tsentral'noe statisticheskoe upravlenie, *Narodnoe khoziaistvo SSSR v 1958 godu* (Moscow: Gosstatizdat, 1959).

A. Troitskii, 'Elektroenergetika problemy i perspektivy', *Planovoe khoziaistvo*, February 1979.

A. Troitskii, 'Novye rubeshi sovetskoi energetiki', *Planovoe khoziaistvo*, December 1976.

US Bureau of the Census, *Historical Statistics of the United States, Colonial Times to 1970*, bicentennial edn (Washington, DC, 1975), pt II.

B. Vainshtein, A. Khaitun and N. Sokolov, 'Effektivnost' neftegazovogo kompleksa zapadnoi Sibirii', *Voprosy ekonomiki*, October 1979.

A. V. Venediktov, *Organizatsiia gosudarstvennoi promyshlennosti v SSSR (Leningrad: Izdatel'stvo Leningradskogo Universiteta, 1957).*

Raymond Vernon (ed.), The Oil Crisis (New York: Norton, 1976).

N. Volkov, 'Struktura vzaimnoi torgovli stran SEV', *Vneshniaia torgovlia*, December 1966.

Dow Votaw, *The Six-Legged Dog, Mattei and ENI, A Study in Power* (Berkeley: University of California Press, 1964).

Zbynek Zeman and Jan Zoubek, *Comecom Oil and Gas* (London: Financial Times, 1977).

A. I. Zubkov, 'SSSR i reshenie toplivno-energeticheskoi i syr'evoi problemy v strankakh SEV', *Istoriia SSSR*, no. 1, 1976, p. 60.

A. G. Zverev, *Finansy i sotsialisticheskoi stroitel'stvo* (Moscow: Gosfinizdat, 1957).

Index